What This Book Is About

Create an Oasis with Greywater **describes how to choose, build, and use 20 types of residential greywater reuse systems in just about any context: urban, rural, or village. It explains how you can put together a simple greywater system in an afternoon for under $40. It also includes information for taking your greywater reuse to the next performance level:**

- how to get clear on your goals, and how various greywater system options can serve you
- the same Site Assessment Form and procedures I use for my design consulting
- common mistakes and how to avoid them
- greywater plumbing principles and procedures in detail
- information on soils and plants, tools and parts
- plenty of our own original design innovations to improve the longevity of greywater systems, make maintenance easier, and reduce environmental impacts
- real life examples of greywater designs for a wide range of contexts

This book offers underlying design principles as well as design specifics. If you run into a situation not specifically covered, there's a good chance you'll be able to use these general principles to figure it out yourself.

Most of the world's aquifers are being pumped faster than replenished, and all reservoirs are slowly diminishing in capacity as they fill with sediment. At the same time, natural surface waters and groundwaters are being degraded by the wastewater continually dumped into them. Greywater reuse enables you personally to do more with the same amount of water and to increase your water security. At the same time, your greywater reuse reduces the problems of supply and pollution for everyone.

Any greywater system will realize some benefits. Obtaining *all* the potential benefits is trickier than it seems. Many pitfalls await the unwary. In the average installation, this book will pay for itself many times over in savings on construction, maintenance, and errors avoided.

Humans enjoying responsible stewardship of their part of the water cycle.

Most of the information otherwise available on greywater comes from vendors. Oasis Design doesn't sell greywater systems, so you don't have to worry that we're steering you toward stuff you don't need. Rather, we make our living by providing information to help people have a higher quality of life with lower environmental impact.

Wishing you the best of luck with your projects,

Art Ludwig   Park[er Abercrombie]   [Michelle H]oward

Contents

> ***What Do C, D, and R Mean?***
> *C: Collection plumbing*
> *D: Distribution plumbing*
> *R: Receiving landscape*
>
> *These are the parts that together make a complete greywater system.*

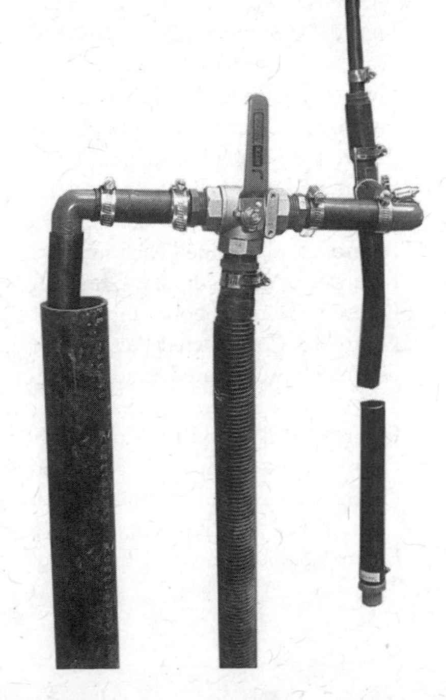

Diversion setup for simple, high-performance Laundry to Landscape system (p. 58)

Figures and Tables

Mulch-filled greywater/rainwater basin nourishe[s] a peach tree in Arizona.
The downspout extension is for rain. The pipe exiting the foundation is shower greywater, which splits between this and another basin via a Branched Drain (outlet in rock work). The tree is planted on an island in the middle to keep the roo[t] crown dry. (In all but the driest climates, rainwate[r] should be movable and divertable away from the greywater basins to avoid overflow or saturation.)

Introduction

On Christmas Eve of the year I turned 15, my present to the family was to move into a decaying tack room in the lower reaches of our backyard. The plumbing consisted of a lone garden faucet outside. If I could deal with a lack of plumbing while backpacking, I reasoned, I could deal with this. The eco-cidal adult establishment thought it needed all that over-engineered infrastructure, but I surely didn't.

To my family's delight, I maintained an independent household based on the faucet and later improvements (wood-heated outdoor bathtub, etc.). In fact, my systems became so refined I didn't leave until I had a wife and a two year old daughter, 14 years later.

During a break in my backyard stay, I bicycled around the world for three years, witnessing myriad philosophies of wastewater management. Later, I studied ecological design at UC Berkeley. My thesis was that *every house should be surrounded by an oasis of biological productivity nourished by the flow of nutrients and water from the home.* Most elements added to water in the home are nutritious for plants. If the small amount of toxins in cleaners could be eliminated, washwater could nurture an edible landscape sustainably.

Greywater reuse follows the same principles that make wild rivers clean…even though they drain many square miles of dirt, worms, and feces. Beneficial bacteria break down nasties into water-soluble plant food, and the plants eat it, leaving pure water. The author is shown here deeply absorbed in his tireless study of this process.

However, I found that cleaners bio*compatible* with plants or soil were not available. But through my research, I learned enough, and befriended enough experts, to assemble a team to create the world's first plant and soil biocompatible cleaners.

Meanwhile, my hometown of Santa Barbara, in an advanced stage of drying up and blowing away due to prolonged drought, became the first locale in the nation to legalize greywater use for irrigation. Consequently, I took a six year side trip to develop, produce, and market Oasis Biocompatible Cleaners, launched in Santa Barbara on Earth Day, 1990.

The amount of time that my staff and I spent answering questions about greywater plumbing quickly became untenable, so I wrote a book about it. I sold the cleaner business in 1996 and have been consulting, writing, and gardening since then. The Oasis greywater books evolved from an incidental to the main part of my business today. Oasis Design has walked central, neutral ground between warring factions during sweeping changes in the emergent greywater industry. Our customers entrust us with their darkest greywater secrets, system manufacturers keep us abreast of their offerings, academics appreciate our research, and, as we have become the world's greywater "information central," regulators seek us out for help writing laws.

People have schlepped greywater around in buckets since time immemorial. However, the modern generation of systems, which attempt automated, efficient delivery, have been in use only a few decades. Legal requirements favor engineering overkill, while the simple and economical methods—the ones that people actually use—remain technically illegal. Happily, visionary jurisdictions such as Arizona, New Mexico, and Texas are starting a strong new trend toward rational regulation. Even so, most government agencies and contractors still know little about practical greywater systems. And, in jurisdictions where practical systems remain illegal, they can't help you anyway. This information void leaves hundreds of thousands of do-it-yourselfers to "reinvent the wheel" with little useful guidance.

Oasis Design's books have filled this void since 1991. This edition of *Create an Oasis with Greywater*—the world's most popular greywater book*—incorporates all this collective wisdom, several of my original designs, all the content from our *Branched Drain Greywater Systems* book, and 50 pages of new material in one comprehensive design and installation manual. (The *Builder's Greywater Guide,* which covers permits, professional installation, and greywater science, remains separate.) With eight times the information in the first edition, this book presents guiding principles for breaking new ground, as well as an expanded range of proven designs and practices, legal and illegal, ancient and experimental.

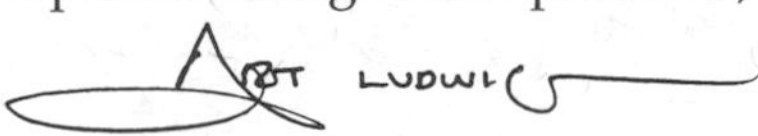

Art Ludwig
San Jose Creek, Santa Barbara, California

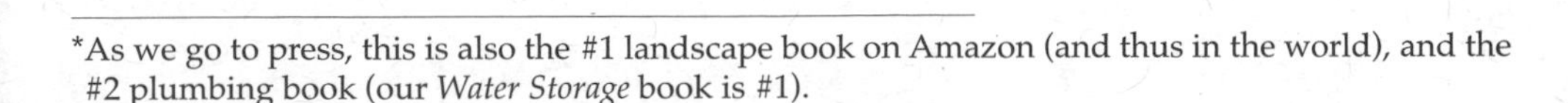

*As we go to press, this is also the #1 landscape book on Amazon (and thus in the world), and the #2 plumbing book (our *Water Storage* book is #1).

Chapter 1: Greywater Basics

First, let's get your feet wet (so to speak)—what is greywater, what can you do with it, why, how, and some greywater lingo.

What Is Greywater?

Any wastewater generated in the home, except water from toilets, is called greywater. Dish, shower, sink, and laundry greywater comprise 50–80% of residential "wastewater." Greywater may be reused for other purposes, especially landscape irrigation.

Toilet-flush water is called blackwater. A few systems that can safely recycle toilet water are included in this book.

Contaminated or difficult-to-handle greywater, such as solids-laden kitchen sink water or water used to launder diapers, I call "dark greywater"; most US regulators consider these blackwater. However, the level of pathogens in even the darkest greywater is a small fraction of that in blackwater.[1]

Wastewater without added solids, such as warm-up water from the hot water faucet, reverse-osmosis purifier drain water, or refrigerator compressor drip, is called clearwater.

Reclaimed water is highly treated mixed municipal greywater and blackwater, usually piped to large-volume users such as golf courses via a separate distribution system. There are serious ecological reservations about this practice, which are outside the scope of this book.

"Any wastewater not generated from toilet flushing" is the official greywater definition commonly used in Europe and Australia.

—Water Environment Research Foundation[41]

What Can You Do with Greywater?

Conventional plumbing systems dispose of greywater via septic tanks or sewers. The many drawbacks of this practice include overloading treatment systems, contaminating natural waters with poorly treated effluent, and high ecological/economic cost.[2]

Instead, you can reuse this water. The most common reuse of greywater is for irrigation—the focus of this book. It can also be cascaded to toilet flushing or laundry. Even a greywater-only disposal system has less negative impact than septic/sewer disposal.

Why Use Greywater?

It is said that there is no such thing as "waste," just misplaced resources. Greywater systems turn "wastewater" and its nutrients into useful resources. Why irrigate with drinking water when most plants thrive on used water containing small bits of compost?

Unlike many ecological stopgap measures, greywater use is part of the fundamental solution to many ecological problems. It will probably remain an essentially unchanged feature of ecological houses in the distant future. The benefits of greywater recycling include:

- **Reduced use of freshwater**—Greywater can replace freshwater for some uses. This saves money and increases the effective water supply, especially in regions where irrigation is needed. Residential water use, on average, is almost evenly split between indoors and outdoors. Most water used indoors can be reused outdoors for irrigation, achieving the same result with less water diverted from nature.
- **Less strain on septic tanks or treatment plants**—Greywater, which comprises the majority of the wastewater stream, contains vastly fewer pathogens than blackwater and 90% less nitrogen (a nutrient that is a problematic water pollutant). Reducing a septic system's flow by getting greywater out greatly extends its service life and capacity. For municipal treatment systems, decreased flow means higher treatment effectiveness and lower costs.
- **More effective purification**—Greywater is purified to a spectacularly high degree in the upper, most biologically active region of the soil. This protects the quality of natural surface and groundwaters. Topsoil is a purification engine many times more powerful than engineered treatment plants, or even in septic systems, which discharge wastewater deeper into the subsoil.[3]
- **Feasibility for sites unsuitable for a septic tank**—For sites with slow soil percolation or other problems, a greywater system can partially or completely substitute for a costly, over-engineered septic system. (In extreme cases this can enable otherwise undevelopable lots to be built on—a double-edged sword environmentally.)

Greywater Lingo, Measurements

Words with squiggly underline you'll find defined in the text and cross-referenced in the Index.

Unfamiliar measurement? See Appendix H: Measurements and Conversions.

- **Reduced use of energy and chemicals**—Due to the reduced amount of freshwater and wastewater that needs pumping and treatment. If you provide your own water or electricity, you'll benefit directly from lessening this burden. Also, processing wastewater in the soil under your fruit trees definitely encourages you to dump fewer toxins down the drain.
- **Groundwater recharge**—Greywater application in excess of plant needs recharges the natural store of water in the ground. Abundant groundwater keeps springs flowing and trees growing in intervals between rains.
- **Plant growth**—Greywater can support a flourishing landscape where irrigation water might otherwise not be available.
- **Reclamation of nutrients**—Loss of nutrients through wastewater disposal in rivers or oceans is a subtle but highly significant form of erosion. Reclaiming otherwise wasted nutrients in greywater helps to maintain the land's fertility.[2,3]
- **Increased awareness of, and sensitivity to, natural cycles**—The greywater user, by having a reason to pay more attention to the annual progression of the seasons, the circulation of water between the Earth and the sky, and the needs of plants, benefits intangibly but greatly by participating directly in the wise husbandry of vital global nutrient and water cycles.
- **Just because**—Greywater is relatively harmless and great fun to experiment with. Moreover, life with alternative waste treatment is less expensive and more interesting.

The competition: A seawater desalination plant *provides some of the world's most costly, energy-intensive freshwater.*

An activated sludge sewage treatment plant *dumps greywater in the ocean after energy- and chemical-intensive treatment.*

When Not to Use Greywater

There are a number of possible reasons *not* to use greywater, or to use it only during certain times of year:

- **Insufficient space**—In some situations, neighbors are too close, or the yard too small or nonexistent. (Though even a tiny suburban lot with lots of people can work, as can indoor greywater reuse. The latter is mostly outside of the scope of this book, but see Appendix F and oasisdesign.net/greywater/indoors).
- **Inaccessible drain pipes**—If all plumbing is entombed under a concrete slab, accessing most of the greywater won't be economical (but you can almost always get at the laundry water, and a combined wastewater reuse system may work).
- **Unsuitable soil**—Soil that is either extremely permeable or impermeable may preclude the use of a greywater system or at least require special adaptations.
- **Unsuitable climate**—In very wet climates, where using greywater for irrigation is of little benefit, better ways to dispose of it may be available. In very cold climates, freezing may prevent the use of a greywater system for part of the year (but see Appendix C).
- **Legality concerns and/or permit hassles**—In much of the industrialized world, the legality of greywater systems is a "grey" area. However, there seems to be movement toward a less paranoid, more realistic official attitude, concurrent with increased experience and improved systems, not to mention water shortages and pollution problems.[4] For example, nearly all of the systems described here are legal in Arizona, New Mexico, and Texas, thanks to laws passed in 2001, 2003, and 2005, respectively. In most of the rest of the US, most practical greywater systems are not allowed. However, authorities generally turn a blind eye toward greywater use even where illegal. The State of California, for example, once published a pamphlet that explained the illegality of greywater use but then explained how to do it—and get a tax credit for it!
- **Health concerns**—The rationale for keeping greywater illegal in many areas is supposed concern for public health. In practice, the health threat from greywater has proven insignificant. No one seems to know of a single documented instance of a person in the US becoming ill from greywater. A quantitative field test by the Department of Water Reclamation in Los Angeles found that greywatered soil did teem with pathogens—but so did the tapwater-irrigated control soil. The study's conclusion: Don't eat dirt, with or without greywater.[5] (Health concerns are considered in detail on p. 18.)
- **Poor cost/benefit ratio**—In some situations, especially when legal requirements mandate a complex system for only a small flow of water, the economic and ecological costs may outweigh the benefits. This is common with professionally installed systems, which are generally much more expensive than owner-installed ones.
- **Inconvenience**—So far, many greywater systems are more expensive or require considerably more user involvement than well-functioning septic or sewer systems.

- **Lowered return flows**—In the Southwest, in particular, laws credit "return flows"—treated wastewater returned to a river—against water withdrawals. That is, the more wastewater a city dumps in a river, the more it can take out. Greywater reuse for irrigation reduces return flows, but it also reduces the amount of freshwater needed for irrigation, and the amount of treatment and pumping. No doubt a complete accounting would find direct greywater reuse to be more pro than con for the overall water/ecology balance.
- **Insufficient combined wastewater flow**—In the unlikely scenario that everyone reused *all* greywater, the reduced flow through municipal sewers designed for higher flow could be insufficient to move toilet solids through.
- **Inappropriate development**—Greywater systems and composting toilets are sometimes pursued by developers whose only interest is developing property that is otherwise unbuildable due to soil in which a septic system isn't allowed.

Greywater Facts and Figures

- *7% of US households reuse greywater*
- *13% of California and Arizona households use greywater for landscape irrigation, with clothes wash water the most utilized source (66%)*
- *Greywater constitutes about 50% of total (69 gallons/person/day) household wastewater generated*
- *An average (2.6 person) household has 90 gallons of greywater/day available for outside use, not including the kitchen sink*
- *A home-owner with a 2,500 ft² house on a ¼ acre lot could irrigate about ½ the yard with greywater if xeriscaping is used.*
- *In Arizona, a two year study on the landscape and plants irrigated with greywater in residential areas revealed that, except for a slight increase in boron, no salts had accumulated in either the plants or the surrounding soil.*

—Water Environment Research Foundation[41]

Elements of a Greywater System

What a greywater system does is collect greywater, then divide and distribute the flow among planted areas. Most of the systems described here are variations on this basic functionality, even those systems designed primarily to get rid of greywater rather than extract value from it. The "hardware" that accomplishes this generally consists of:

1 **Greywater source(s)**—E.g., washing machine, shower, bathtub, and/or sinks.
2. **Collection plumbing**—Pipes that transport greywater from the house to one or more points just outside the house.
3. **Surge tank, filter, and pump**—Optional elements that add complexity and cost but can make the distribution plumbing's job easier, especially for large flows.
4. **Distribution plumbing**—Plumbing that transports greywater through the yard and divides it among plants.
5. **Receiving landscape**—Soil, roots, plants, and mulch basins that contain, cover, purify, and use the greywater.
6. **People**—Those who design, make, and maintain the system, generate the greywater, tend the garden, and eat the fruits. People are a critical, but often overlooked, component of the overall system.

The drawing below shows a few simple greywater systems with parts labeled. Check it out, then roll up your sleeves, sharpen your pencil, and proceed to Chapter 2.

FIGURE 1.1: ELEMENTS OF GREYWATER SYSTEMS
(KEY IS LIST ABOVE)

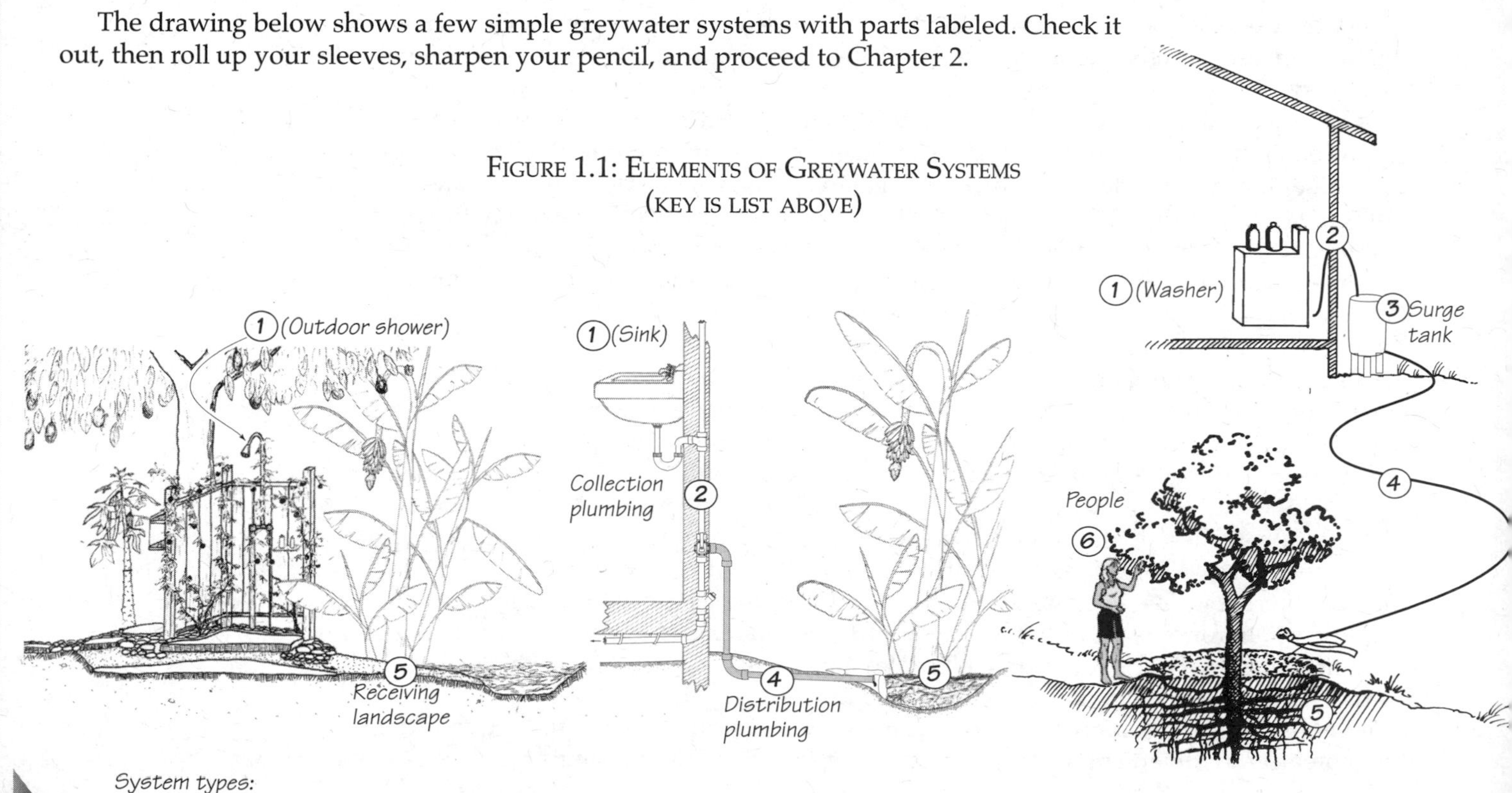

Chapter 2: Goals and Context

If you just want to start building a simple system, you can skip ahead to the System Selection Chart, p. 51. However, even the simplest system will turn out better if you read straight through...

This book covers the territory of just about every possible greywater system. Here is the general path to the right system for you, around which this book is organized:

- get clear on your goals and assumptions
- assess your context
- choose the most appropriate greywater distribution system for your context
- design the connections to rainwater harvesting, landscape, etc.
- design the greywater collection plumbing (if needed)
- build the system from top to bottom, following the flow from fixtures to receiving landscape
- use it, maintain it, troubleshoot it

In this chapter, you'll get clear on your goals and assumptions, and then assess your context—in other words, what you want, and where you are.

Get Clear on Your Goals

Unless you are totally shooting from the hip, I strongly suggest you photocopy or clip out the ***Site Assessment Form*** *(Appendix A), and fill it with your goals and site data as you proceed through this chapter. There's an example of site assessment info on the next page.*

This form is a great way to organize your thoughts. It's the same form I use when designing a system for a client. (You can download an editable, printable version from our website, oasisdesign.net/design/consult/checklist.htm.)

The foundation of ecological design is getting clear on your goals and context. These affect every aspect of the design in a major way. In conventional construction, this step often doesn't exist, because it doesn't need to; conventions and codes are understood to guide the project. In ecological design, you're inventing a new system rather than fitting into an existing one. It is useful and necessary to re-examine your assumptions each time. Start by asking:

- **What is your guiding philosophy and aesthetic?**—Is the project in an exclusive gated subdivision, or a mountain shack in a hippie commune?
- **How are you willing to change your lifestyle?**—Lifestyle change is far and away the best way to conserve resources. For example, if you decide you don't really need long, high flow showers and a vast ornamental lawn, that decision will save most of the water and energy right off the bat. Then you can install a simpler, cheaper greywater system for the water you're still using. If you can choose between reducing and reusing, reducing is the priority. (For more on this topic, see our booklet *Principles of Ecological Design.*)
- **What perfection standard are you working to?**—Do you want aesthetics, efficiency, etc. to be 70, 90, 95, or 98% perfect? This is a cultural and personal preference. Each standard has its charm—and cost. Each jump to the next highest of these standards roughly *doubles* the amount of money, time, and materials required for the job, and the negative environmental impact as well. Perfection standard is a huge variable but is rarely discussed explicitly.
- **Are you trying to reuse your greywater, or just get rid of it safely?**—Responsible disposal and reuse are both beneficial. In either case, don't forget to explore conservation first.
- **How much effort do you want to put into increasing reuse efficiency?**—It is quite a bit more effort to ensure that greywater use will actually lower your freshwater use.
- **What are your other greywater system goals?**—Improve sanitation? Reduce pollution? Save a failing septic? Save money? Feel good? Make a demonstration? (Ideally, demonstrations should justify themselves without resort to the theory that inspiring other people to "do the right thing" makes up for waste on the demonstration itself.)
- **Do you want a particular economic payback time?**—Greywater system economics are best when you save septic pumping and replacement, and for do-it-yourself systems.
- **What are your landscape goals?**—Beauty? Food production? Erosion control? Slope stabilization? Firebreak? Privacy screen? Outdoor living oasis?

Give it some thought, and write your goals down. If you're working with others, include them in the discussion.

Assess Your Context

Now we'll turn to your specific context—your site, greywater sources, irrigation need, percolation rate, climate, etc.

Greywater Systems Are Very Context Dependent

The one general principle in greywater system design: ***There are no general principles.*** Greywater systems are very context dependent. That is, a small change in your context can mean drastic change in your system selection and design. Your particular path through this territory (and this book) depends on your context. This book is organized around the following middle path:

A carefully considered and optimized, do-it-yourself, residential, retrofit system for gravity flow irrigation, not too ritzy, built without a permit, in a moderate climate, with a plumber involved in the collection plumbing, and a septic or sewer backup.

Multiple side trips or shortcuts from this main path may apply to you—on the next page we'll see which ones.

Does Greywater Matter?

Viewed from any single, narrow perspective, greywater systems don't look that important. A low flow showerhead can save water with less effort. A septic system can treat greywater almost as well.

But when you look at the whole picture—how everything connects—the keystone importance of greywater is revealed.

Ecological systems design is all about context, and integration between systems.

Ecological systems—rainwater harvesting, runoff management, passive solar, composting toilets, edible landscaping—all of these are more context sensitive than their counterparts in conventional practice.

And greywater systems are more context sensitive than any other manmade ecological system, and more connected to more other systems.

Get the greywater just right, and you've got most of the whole package right, and that's what matters.

Site Assessment Example

(This is the site assessment for Real World Example #1, Chapter 12.)

Greywater system goals: Irrigate large garden without excessive water use, dispose of water safely, reduce ocean pollution, and demonstration for State of California greywater study.

Landscape goals: Beauty, food production, privacy screen, outdoor living.

Perfection standard: Moderately high. The house and yard could be in *Sunset Magazine*, somewhat on the rustic end.

Location: City of Santa Barbara, latitude 34°, altitude 150' *(45 m)*, ¼ acre lot *(1,000 m²)*, 1950s suburb.

Water system: City water meter with tiered rates that climb steeply with water use. City water deliveries were down 70% in a prior drought.

Rainwater and runoff: All absorbed onsite due to thick mulch layer, basins, and moderate slope. There is no surface water nearby. Groundwater is 50' *(15 m)* down.

Existing wastewater treatment facilities: Sewer, which ultimately dumps in the ocean.

Population of water users: 3 average, 20 peak, 1 minimum.

Landscape and irrigation: Native vegetation was coastal scrub. The whole lot is gardened intensively with a mix of fruit trees, vegetables, and flowers, all underlain with thick wood chip mulch. The garden is low water use for how extensive, beautiful, and productive it is.

Slope: 2% to the NE, away from the house. The whole yard is below the water sources.

Soil and groundwater: Soil is rich and loamy, gardened organically for 50 years, excellent perk, and easy to dig. Biocompatible cleaners used exclusively.

Climate: Rainfall = 4–44"/year, 18"/year average,[m] max evapotranspiration = 1½"/week between November and March, regularly 6 months with zero rainfall, few to no killing frosts.

Forces of nature: Severe recurring drought.

Site map, system elevations: see Real World Example #1, p. 111.

People: The owners did most of the work on the system, with help from a plumber on the collection plumbing and yours truly on the distribution plumbing. Owners are enthusiastic gardeners and conservers of water. Original users were very keen on maintaining the system, the current owners less so.

Regulatory climate: Permitted under California greywater law as part of a state study.

Economics: The owners didn't care too much about the economic payback time. The sewer plant savings don't accrue to them, only the savings on water. These will pay back the cost of the system in 10–20 years (half that if you don't count added costs from measuring the water and complying with unnecessary legal requirements). As a participant in a landmark study, this site had significant political and scientific value.

System used: Branched Drain with dipper, to sub-mulch emitters.

[m]***Metric:*** *10–110 cm/year, 45 cm/year average.*

Side Trips and Shortcuts That May Apply to You

Check this section to identify which kind of greywater user you are, so you'll know when I'm talking to YOU—pay attention!—and when to skim or skip ahead. In addition, most sections start with an indication of whom they apply to.

Quick and Dirty *(A shortcut compared to carefully considered and optimized)*

Greywater systems can be made with a great deal of care and a close eye to optimizing the outcome; or, they can be slapped together without much thought or trouble—as the vast majority are. Greywater work is pretty forgiving; no matter how it gets done, you'll realize a portion of the possible benefits. Either approach (or anything between) is valid. If you just want to slap together a quick, super-simple system and don't care whether it is 50 or 90% perfect, you can skim or skip most of this book, focusing on Chapter 7: Simple, Easy Greywater Systems.

Disposal Only *(A shortcut compared to reuse for irrigation)*

If your system is to be for disposal only (for example, to take the pressure off a septic system), then the large amount of information on optimizing irrigation efficiency will only concern you inasmuch as it helps you spread greywater over a wide area and avoid overloading the soil with water.

Multi-Unit, Institutional, Industrial, High Flow *(A longer journey than residential)*

Residential systems have two defining characteristics: The same people who generate the greywater inhabit the landscape it waters, and the flow is low. Different laws generally regulate high flow (multi-unit, institutional, and industrial) systems. They usually require filtration and pumps, as well as professional installation.

The principles described in this book are relevant at any scale. A few of the systems described can be scaled up to serve 100 or so people (see System Selection Chart, "Suitable for large flows" column). For more information on medium-sized systems, see our *Builder's Greywater Guide.*[6] Giant centralized systems are beyond the scope of our publications.[7]

New Construction *(Can take less or more time than a retrofit, and yield more benefit)*

If your system is to be retrofit to an existing building, you're pretty much stuck working with what you've got. But if it's for new construction, you have the opportunity to make everything better integrated—a much more challenging and potentially rewarding task. You have to coordinate the efforts of the architect, plumber, engineer, and landscaper, even if you are all these people yourself (see Figure 3.1).

Greywater use is best incorporated during the design phase. Like passive solar, it is a primary design consideration. It affects the location of the house and every other water feature. The higher the greywater sources above the irrigated area, the better. The more accessible the drainpipes, the better.

Elevation relationships between water features are critical and are difficult to change once the house is built. Don't let your plumber squander fall (the vertical height between parts of the system)—make sure the pipes come out as high as possible (see Squander No Fall, p. 22).

Although legality is virtually never an issue for retrofits, it is virtually always an issue for permitted new construction or remodeling.

Permitted *(A longer, more exasperating path that may not go where you want)*

You usually must permit a system when it is part of permitted new construction or remodeling (unless your system is in Arizona, New Mexico, or Texas—then you're home free).

If permits are involved, you'll need to deal with impractical requirements. To get through this maze, use our *Builder's Greywater Guide,* written expressly for folks dealing with permits. You may choose to arrange plumbing conventionally first, to pass inspection, and then add greywater plumbing later. If you keep blackwater and greywater lines separate, installing a greywater system later is easy. Arranging things to pass inspection, and plumbing drain lines to work in septic/sewer as well as greywater modes, requires the services of an especially skilled, cooperative plumber.

Most reasonable greywater systems, including all versions of the Branched Drain and Laundry to Landscape, are allowed in Arizona, New Mexico, and Texas. The Uniform

Plumbing Code (UPC) is the model code for most Western states. Many have adopted its greywater appendix, which was based on the greywater appendix of the California Plumbing Code (CPC). But the UPC and CPC are so maddening and unreasonable that the vast majority of greywater users subject to it only get a permit if they absolutely must (see oasisdesign.net/greywater/law).

Professionally Installed *(In some ways a longer journey than do-it-yourself, but then you have help)*

If you want a professionally produced, turnkey system, the first hurdle is to find a professional greywater installer—a rare breed (see oasisdesign.net/links). The second hurdle is that professionals tend to make the process more complicated, and certainly more expensive. If you want to do as little work as possible yourself, the most realistic approach is to act as a contractor and hire specialists to help with the tasks in Figure 3.1. The more you can do yourself, the better—except the collection plumbing.

Do-It-Yourself Collection Plumbing *(A false shortcut compared to professional collection plumbing installation)*

Remember the no-general-principles principle? Whereas most greywater work is better the more of it you do yourself, collection plumbing generally turns out better the *less* of it you do yourself. Reading Chapter 4: Greywater Collection Plumbing will set you up either way.

Pumped *(Longer, more convoluted journey than gravity flow)*

You're best off not pumping unless you have new construction, high flows, or nowhere to go but uphill—see Appendix E.

Extremely Cold, Wet, Dry, Etc. *(More involved than moderate climate journey)*

For extreme cold, read Appendix C. For extreme wet, see What to Do with Greywater When You Don't Need It, Chapter 5. A variety of other extremes are covered in Chapter 12: Real World Examples.

Very Low or No Perk Soil *(More extensive and expensive than high perk path)*

If your greywater loading rate is high and water infiltrates slowly into your soil (low perk), you may end up with standing water. If the soil space doesn't have a chance to fill with some air at least once every 24 hours, it may get anaerobic and smelly. If spreading the greywater more evenly over a wider area can't prevent this condition, consider a lined Constructed Wetland, or an unlined bog with wetland plants, or a Solar Greywater Greenhouse, Chapter 8.

No Plants or Soil *(Totally different path)*

If the system is only for indoor reuse—for example, flushing toilets with greywater—and involves no plants or soil, purification will be entirely artificial. Such systems are not generally economical or ecological on a residential scale, so this book doesn't cover them. However, we list some suppliers at oasisdesign.net/greywater/references.htm and some novel systems at oasisdesign.net/greywater/indoors.

Failing Leachfield *(Strong incentive to get moving!)*

A failing septic leachfield[8] can be restored to function by lowering or eliminating the flow of greywater through it. The first priority is to divert the laundry water, as lint tends to pass through the tank and *papier-mâché* the leachfield surface shut. The next priority is diverting the rest of the greywater.

It is worth noting that savings on septic tank pumping and leachfield replacement will pay off almost any greywater system. And all you have to do is get the greywater out of the septic system—no optimization or other complication is necessary.

High Water Table *(More involved journey)*

If you have a high water table and/or exceedingly fast perk, see Constructed Wetlands and Solar Greywater Greenhouse, Chapter 8.

No Septic/Sewer *(Simpler than greywater with septic/sewer backup)*

If radical simplicity draws you to the greywater system/composting toilet lifestyle, see Radical Plumbing, Chapter 4.

System in Non-Industrialized World *(Simpler, different path)*

The hardware available in non-industrialized countries is different, and it is cheaper to custom-make items such as 3-way valves than to buy them. Solutions that are labor intensive in use are more practical. See Radical Plumbing, Chapter 4, and Appendix D.

Now that you've got an idea of what kind of greywater user you are, it will be easier to recognize what applies to you and what doesn't as you press onward.

If you start to feel overwhelmed with information, remember that you've got the option to lower the bar quite a lot—just make a quick and dirty system—and things will still turn out okay. Take further comfort in knowing that even the worst greywater disasters don't amount to much as disasters go. You can get in a lot more trouble with wood stove installations, electrical work, blackwater systems, or any number of other do-it-yourself projects.

Assess Your Site

Continue p. 18

You can skim this if you don't care about optimizing your system.

Filling out the aforementioned Site Assessment Form (Appendix A) can facilitate site assessment. A site map is another great aid for designing a system. A map with a scale of ⅛" = 1' *(1:100)* with 1' *(25 cm)* contours is ideal. If such a map is not available, make a sketch showing the house, boundaries, irrigated areas, natural water features, utility lines, routes of plumbing under the house, etc.—all the features listed on the Site Assessment Form. You'll want a few copies of your site map, or tracing paper to lay over it. There are examples of site maps on p. 13, 40, 113, and 116.

Assess Your Water Resources

The first question for any human habitation is, "Where does the water come from?" Ideally, for minimum impact and longterm sustainability, **rain should be the primary water source, followed by greywater, followed by gravity flow spring or creek water, followed by well or municipal water.**[9] Wherever the water comes from, note its characteristics and limitations on the Site Assessment Form. Water storage[10] is also key—note this as well.

Evaluate Conservation Options

If you aim to live ecologically, thoroughly explore conservation options before turning to reuse. Because of the unavoidable inefficiencies inherent in greywater systems, *the gain from reducing consumption with water-conserving fixtures is always greater than that from reusing the water.*

First, reduce water use with efficient fixtures. In this "forever hot" bathtub in Mexico, water is heated from below by gas burners. The sides of the tub are insulated. There is no need to add more water just to heat the tub. Also, such a tub can be filled with solar pre-heated water, then brought to temperature, which is more efficient—and potentially eliminates the need for any other water heater.

Assess Existing Wastewater Treatment Facilities

The state, capacity, and location of other wastewater treatment systems influence greywater system design. Locate these on your site map and describe them in the Site Assessment Form. If you're on a septic system, you'll easily pay back your greywater system by septic savings. If you're on a sewer line, you won't benefit economically from diverting the greywater flow, but your community will.

Assess Your Greywater Sources

For any system in any context, I recommend doing at least a rough reality check to make sure the quantities of output and intended use are in the same ballpark.

It is commonly assumed that however much greywater there is and however much irrigation is needed, they'll magically match up. Even if you don't want to reuse the greywater, it's good to have an idea of how much you've got in order to know if your disposal/treatment area is big enough. For most purposes, you can just take the home's population and multiply it by an approximate daily use figure, as follows:

Homes without water-conserving fixtures generate about 55 gal/person/day. Efficient fixtures cut this daily output to 40 gal/person/day, or

TABLE 2.1: GREYWATER SOURCES, QUALITIES, AND QUANTITIES

Source *Ease of replumbing*	Quality *Ways to improve quality*	Quantity	*Metric Quantity*
Sources with their own pumps			
Washing machine *Easiest collection plumbing. Getting washer lint out of septic leachlines greatly extends their life.*	**Good.** Medium concentration of soaps, lint. Diapers can dramatically increase pathogen level. *Can be improved to **excellent** by using biocompatible cleaners.*	**Large:** 30–50 gal / load (10 gal for front loader). 1½ loads / week / adult, 2½ / child. *85–100 gal/ person/week.*	*320–380 L/ person/week*
Automatic dishwasher *May be easily replumbed by a do-it-yourselfer.*	**Poor.** Low to high quantity of solids, depending upon degree of pre-rinsing. High salt and pH from conventional automatic dishwashing compounds; alternative cleaners don't clean well.	**Small:** 5–10 gal / load. *7 gal/person/week, average.*	*20–40 L/load*
Gravity flow sources			
Shower *Requires professional replumbing. May be impossible with slab foundation.*	**Excellent.** Minimal concentration of soap and shampoo is of little concern. Contains pump-snarling hair. *Use the least amount of soap and shampoo necessary. Use liquid soap to reduce sodium.*	**Large:** 20 gal / person / day for high flow shower; 10 for low flow. *70–100 gal/person/week.*	*76 L/person/ day, 38 L/ person/day, 260–380 L/ person/week*
Tub *Requires professional replumbing. May be impossible with slab foundation.*	**Excellent.** Same desirable qualities as shower, only more so.	**Variable:** 40 gal / adult bath, 25 gal / kid bath. *Use is highly variable.*	*150 L/adult, 95 L/kid*
Bathroom sink *Requires professional replumbing.*	**Good.** Concentration of soap, shaving cream, and toothpaste can be high. *Use liquid soap. Exercise discretion in choice and quantity of other products.*	**Small:** 1–5 gal / person / day. *7–35 gal/person/week.*	*4–20 L/person/ day, 26–130 L/ person/week*
Kitchen sink *Requires professional replumbing.*	**Good** but **problematic** in delicate systems. High in nutrients, but also in solids, grease, and soap. *Despite low pathogens, many authorities consider kitchen sink water "blackwater" not worth trying to reuse. I like it, due to its nutrient value. It can be a design problem for systems, but is not a problem in soil. One workaround is to plumb only the rinse side of a double sink to the greywater system. Meat eaters can add a grease trap.*	**Large:** 5–15 gal / person / day. *35–105 gal/person/ week.*	*20–55 L/ person/day*
Reverse-osmosis water purifier wastewater *May be easily replumbed by a do-it-yourselfer.*	**Excellent.** "Clearwater" with no suspended solids. Contains 25% more of the same dissolved solids as tapwater.	**Medium:** 3–5 gal / gal drinking water used. *13–21 gal/person/week.*	*50–80 L/ person/week*
Water softener backwash *May be easily replumbed by a do-it-yourselfer.*	**Very bad.** Water softener backwash is extremely high in salt (sodium chloride), harmful for plants. *Use of potassium chloride salt[s1] instead of sodium chloride salt can raise quality to **bad,** but is still more of a disposal problem than a reuse opportunity.*	**Small:** 5% of indoor water use.	
Softened water	**Poor.** Softened water contains salt (sodium chloride), harmful for plants. *Potassium chloride[s1] can be used instead, raising its quality to **okay.** The best thing is to disconnect the water softener.*	**All greywater** if softener is in use.	
Toilet water *Requires professional replumbing. May be impossible with slab foundation.*	**Very bad.** High pathogens, suspended solids, and salt. Toilet water is blackwater, inappropriate for reuse in an ordinary greywater system. *In a system designed to address the solids and health issues, toilet water is **very good** (see System Selection Chart).*	**Medium:** 5–8 gal / person / day low flow; three times that for high flow. *60–135 gal/person/week.*	*20–30 L/ person/day, 230–510 L/ person/week*

Notes: Leaks can add 5% or more to water use. Good strainers on drains can dramatically reduce solids.

enough to water four mature fruit trees or a dozen shrubs in an average climate.[11] Extreme conservation habits can cut production to 5 gal/person/day. Households that haul water by hand generate 5–10 gal/person/day of greywater.[m]

To estimate the flow from a single fixture or a subset of fixtures (such as a washing machine, or a shower and bathroom sink), or to do a more detailed analysis, use Table 2.1 and the worksheet in the Site Assessment Form.

You can also use Table 2.1 to get a more precise idea of the total amount of greywater generated. However, even if you have an engineering mind, let go of the possibility of a truly precision engineered system. Greywater volume and irrigation demand vary independently over such a huge range that one can only approximate a match between available greywater and irrigation need.

On your site map, note where the greywater sources are, and (if they are not combined) how much of the total flow comes from each set of sources (Table 2.2). Group fixtures that share a common drain into subtotals (for example, the bathroom sink, tub, and shower often share a drain). Note where their plumbing goes. Consider whether it is appropriate to move any greywater sources (such as a washing machine), add an outdoor shower, plant a tree, or take any other action to improve the fit of water sources to irrigation need. The places where greywater is generated should ideally be high on the property, so you can irrigate more area by gravity. If the greywater sources are all in the lowest corner of the property, irrigation will be concentrated in a small, oversaturated area.

Note variations in the population of water users on the Site Assessment Form. If everyone goes away in the dry season for an extended time, this obviously affects the system's design; you won't want to set up water-loving plants to be irrigated by greywater only.

Gal per Day or per Week?

I use both. gpd figures are more common in other applications. However, some greywater sources may be used less than once per day (bath tub, washer), evapotranspiration changes daily, and soil stores irrigation moisture for about a week. Thus, I think gpw is the most useful greywater design time unit.

TABLE 2.2: EXAMPLE OF WASTEWATER AMOUNTS

Washer: 5 loads/wk × 32 gal/load	=	160 gpw	*600 lpw*
Reverse-osmosis water purifier wastewater (clearwater): ½ gal/day/person × 4 gal rejected/gal used × 4 people × 7 days	=	56 gpw	*210 lpw*
Kitchen sink: 3 gal/day/person × 4 people × 7 days	=	84 gpw	*320 lpw*
Shower: 1 shower/day/person × 8 min × 2 gpm × 4 people × 7 days	=	448 gpw	*1,700 lpw*
Bathroom sink: 2 gal/day/person × 4 people × 7 days	=	56 gpw	*210 lpw*
Tub: 2 baths/wk × 30 gal	=	60 gpw	*230 lpw*
Reusable subtotal	=	864 gpw	*3,270 lpw*
Toilet: 3 flushes/day/person × 1.6 gal/flush × 4 people × 7 days	=	126 gpw	*477 lpw*
Total indoor water use	=	990 gpw	*3,747 lpw*

Call it 1,000 gal for two adults with two kids; this is 35 gal/person/day indoors, below the 50 gal/person/day average for households with piped-in water, and above the 10 gpd average for households without.

Check the Slopes and Elevations

This step is crucial for collection plumbing and gravity powered greywater distribution, especially Branched Drain systems, where slopes must be accurately measured. For Greywater Furrow Irrigation, gauging slope mostly by eye seems to work. For pumped systems, elevations aren't so critical.

Check the slope from beginning to end of the intended system. You need to establish that there is enough fall to get from the greywater sources to destination plants with a constant downhill slope (generally ¼" per foot, or 2%). In the landscape the pipes should ideally still be at surface level, and in no case more than 1' below ground. For proper flow, drainpipes must inexorably get lower with distance from the source, and also with the negotiation of each obstacle, change of pipe size, etc.—and the pipe elevation can't ever go back up again.

Help! Too Many Numbers!

Don't worry if you get them or not. The vast majority of greywater systems are made without any calculations at all—and most still work.

Relax. If you get even a vague sense of the relationships from these pages you'll be way ahead of the curve.

[m]*Metric:* *Without conservation, 210 L/person/day; efficient fixtures, 150 L/person/day; extreme conservation, 20 L/person/day; hand carried, 20–40 L/person/day.*

Fall is almost always a limiting factor for collection plumbing and gravity distribution plumbing (see Squander No Fall, Chapter 4).

If you're fortunate enough to have a 1' *(25 cm)* contour map of your site, use it to check elevations. (A contour map has lines on it showing where the land surface is at the same elevation. On a 1' map, there are lines for each foot. For an example, see Figure 12.1.) If you don't have such a map, measure the elevations where the system is going to be installed. You can do this with a $3,000 transit, but a $5 water level works well enough (see Appendix B: Measuring Elevation and Slope).

Note these elevations on your site map. From slope measurements, you can determine which sources can water which plants. Remember that you actually need a bit more than 2% slope from end to end, to allow for obstacles and other "fall eaters." Particularly for a Branched Drain system with little slope, it is helpful to make a drawing showing the elevations in section view from one end to the other of the system's flattest branch.

Check the Soil Perk

You should have at least an idea of the perk rate for any system. Quantitative measurement is necessary if a lot of greywater is going into a small area, or the perk appears to be low, or the soil is clayey.

The rate at which water absorbs into the soil—the percolation or "perk" rate—is an important variable in greywater system design. High clay content generally means slow perk, which can get dramatically slower if you add salt-laden greywater. Rock can have slow or no perk. Very slow perk can lead to standing water and noxious smells.

Sand is high perk, gravel more so. Very fast perk can lead to groundwater contamination, though this is unlikely if there is a dense network of plant roots.

Fissured limestone and lava tubes can be very problematic for wastewater management, as they form natural pipes that enable wastewater to bypass soil purification and flow directly into the groundwater at warp speed.

It is vital to know about the perk where the greywater is going before committing to a design. I once embarrassed myself by building a system for a neighbor, based on an assumption from my prior experience that the soil around here is very free-draining. Well, the *soil* is free-draining, but where I dug the infiltration basin turned out to be *subsoil* from the house's excavation, and it hardly drained at all.

If you're putting a small amount of greywater in a big area, the only perk issue would be a really, really slow perk rate. If you have enough experience with your soil—digging holes to various depths, perhaps filling them with water, observing how water absorbs into the ground—to know it perks reasonably fast, you don't need a perk test.

But if you are applying a large amount of greywater, or have a small area, or are applying greywater subsurface, a perk test is critical. Based on the result, you might change the size of the system by a factor of 4 or more, or even decide that a greywater system is infeasible.

A professional (read: expensive) perk test is required for installing most septic systems. These tests are usually done with a truck-mounted borer to a depth that is irrelevant to greywater systems. If you already have one of these tests in hand and it shows that the perk rate is okay several feet down, it is surely fine at greywater depth. But a shallow, do-it-yourself perk test that takes an hour gives you more relevant information (see sidebar).

How to Measure Perk

Dig holes to greywater discharge depth (usually 6–12", 15–30 cm), in the locations you wish to discharge it. Place stakes into the bottoms firmly. Mark reference elevations on the stakes and fill the holes with water two to four times, measuring each time how far the water drops in a given number of minutes. When it drops the same amount in the same amount of time, a couple of times in a row, that's your perk measurement (dry soil perks much faster—saturated perk is what matters for sizing the system). Convert the numbers to perk rate in min/in or cm (3" in 60 min is 20 min/in, for example), then see Table 2.3 to see how much area you need to treat that much water.

Measuring perk

If it takes hours for the water to drop at all, or the water vanishes as fast as you can fill the hole, or if the holes fill with water by themselves, you've got a problem! If the perk is super slow, consider a Constructed Wetland. If it is super fast, add compost, mulch, and plants that will develop a dense network of roots to help purify greywater on its short journey through the soil.

Assess Your Treatment/Disposal Area

Skip to the next section if you want to irrigate.

If you only want to dispose of greywater, you can differentiate prospective treatment areas on your site map by indicating how much greywater they can handle. In this case, the only concern is to not overload the soil or sensitive plants (see Table 2.3). Areas with higher perk rate and areas with more wind and sun can take more water. You don't want to apply too much water near the foundation of the house.[12]

These rates are several times the amount needed for irrigation. Disposal rate—the Long Term Acceptance Rate (LTAR)—is limited by soil perk rather than plant transpiration. With a high application rate, the majority of the treated water ends up percolating down to groundwater rather than transpiring to the sky.

TABLE 2.3: DISPOSAL LOADING RATES

Soil infiltration rate, min/in	**Loading rate gal/day/ft²**	**Area needed ft²/gal/day**	*Soil infiltration rate, min/cm*	*Loading rate m³/day/m²*	*Area needed m²/m³/day*
0–30	2.5	0.4	*0–75*	*.1*	*10*
40–45	1.5	0.7	*100–110*	*.06*	*20*
45–60	1.0	1.0	*110–150*	*.04*	*25*
60–120	0.5	2.0	*150–300*	*.02*	*50*

These rates are conservative, especially for Mulch Basins. Can be up to 3–5 gal / day / ft² for greywater, up to 10 for secondary treated effluent.

Assess Your Irrigation Need

Skip this section if you only want to dispose of greywater.

Note the vegetation types and the extent of irrigated areas and trees on your site map, indicating areas of plants that like wetter conditions by different colors or shades, or you can simply make a list of irrigated areas and how much water you think they want—both options are shown in Table 2.4.

Again, if your aim is to live ecologically, you should thoroughly explore conservation options in the landscape before turning to reuse to feed water-hog plants.

Estimating irrigation demand is an inexact science. Even getting within a factor of 2 of the real irrigation demand is an ambitious goal. As a rough rule of thumb, figure that **your plants need about ½ gal per week of water for each square foot of plant** *(20 L/week/m²)*.

Double this figure for desert, and halve it for a cool, humid climate (see Table 2.5). You can also double or halve this figure again for plants that use more or less water. For trees, use the same formula (or calculate the area under the canopy and multiply it by the gallons per square foot per week*).

This rule of thumb above does not account for variations in evapotranspiration (ET) rate by season, climate, plant type, seasonal loss of leaves, rainfall, or irrigation efficiency. It does,

TABLE 2.4: EXAMPLE OF IRRIGATION NEEDS ASSESSMENT

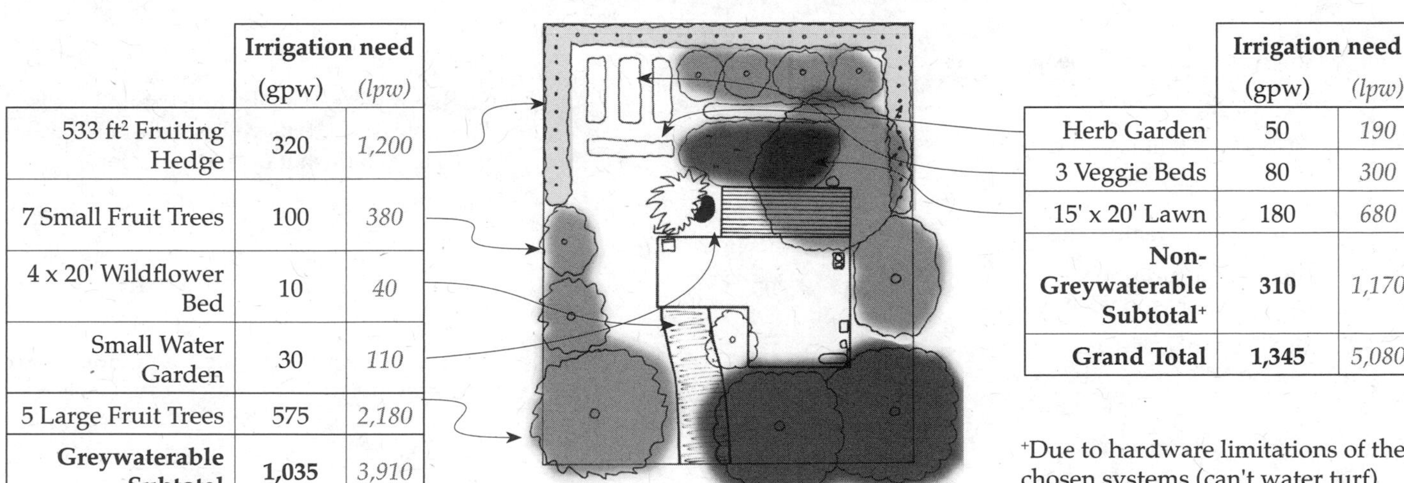

	Irrigation need (gpw)	*(lpw)*
533 ft² Fruiting Hedge	320	*1,200*
7 Small Fruit Trees	100	*380*
4 x 20' Wildflower Bed	10	*40*
Small Water Garden	30	*110*
5 Large Fruit Trees	575	*2,180*
Greywaterable Subtotal	**1,035**	*3,910*

	Irrigation need (gpw)	*(lpw)*
Herb Garden	50	*190*
3 Veggie Beds	80	*300*
15' x 20' Lawn	180	*680*
Non-Greywaterable Subtotal⁺	**310**	*1,170*
Grand Total	**1,345**	*5,080*

This yard belongs to four people. This is 48 gal / person / day outdoors, slightly below average. These numbers are generated with the ½ gal / ft² / wk shortcut. Use 3' *(1 m)* as the smallest diameter for newly planted trees—they like extra water.

⁺Due to hardware limitations of the chosen systems (can't water turf) and health concerns (inadvisable to water veggies).

**To get the area, take half the diameter of the tree canopy, square it, and multiply it by 3.14:* $a = \pi r^2$.

Evapotranspiration (ET) = evaporation from plants plus transpiration through plant leaves. ET increases with temperature, low humidity, and especially wind. These are peak values; generally you want the greywater to cover less (see Choose the Proportion of Irrigation to Meet with Greywater, Chapter 5).

Table 2.5: Peak Evapotranspiration (ET) Values by Climate

	in/wk	*mm/wk*
Cool humid	0.7–1.0	*18–25*
Cool dry	1.0–1.4	*25–35*
Warm humid	1.0–1.4	*25–35*
Warm dry	1.4–1.8	*35–46*
Hot humid	1.4–2.0	*35–50*
Hot dry	2.0–3.2	*50–81*

Cool = under 70° average midsummer high
Warm = 70°–90° average midsummer high
Hot = over 90°F average midsummer high
Humid = over 50% average midsummer relative humidity
Dry = under 50% average midsummer relative humidity

Table 2.6: Sample Greywater Production/Irrigation Need Reality Check (for real world example #2)

(Bold numbers are inputs; the others are calculated)

Greywater sources reality check	gallons			*cubic meters*			
	day	week	month	*day*	*week*	*month*	
Total wastewater calculated	89	**622**	2,674	*0.33*	**2.34**	*10.05*	
Check: Average consumption on wet season water bills	100	698	**3,000**	*0.37*	*2.62*	**11.28**	This checks out; wastewater should be close to but less than wet season water bills.
Check: Wastewater/person	25	174		*0.09*	*0.66*		This is pretty conservative (half of US average) but entirely plausible for a "green" household with low flow fixtures.
Greywater	78	544		*0.29*	*2.04*		
Greywater/person (4 people)	19	136		*0.07*	*0.51*		
Check: % that is greywater		87%			*87%*		High but plausible. The toilet at this site is low flow, and "If it's yellow let it mellow, if it's brown flush it down" philosophy prevails.
Irrigation need reality check							
Average consumption on dry season water bills		2,093	**9,000**		*7.87*	*33.84*	
Check: Difference between wet/dry season bills (presumed to be irrigation)		1,395	6,000		*5.25*	*22.56*	US average is 50% of water use outdoors, so ⅔ during irrigation season is reasonable.
Check: Irrigation as a % of dry season use		67%			*67%*		
Calculated consumption of area to be greywatered		482			*1.81*		
Check: Greywatered area % of total irrigation consumption		35%			*35%*		This looks reasonable, comparing the total irrigated area on the site map for this system with the greywatered area.

however, land you more or less where you'd end up if you *did* consider all those things. If you crave more precision, have extreme conditions, or have a bigger than residential-sized area to irrigate, it may make sense to try a more rigorous analysis, which is covered in our *Builder's Greywater Guide*, Evapotranspiration Formula.

Greywater Sources and Irrigation Need Reality Check

Look over your past water bills or meter records. Your total water consumption for the months when irrigation was not needed should be a good approximation of your *year-round* indoor consumption (greywater and toilet water). Subtract this from your water consumption during months when irrigation is needed, and the number you get should be your irrigation use. These numbers should jibe with your irrigation estimates (see Table 2.6). Leakage and non-irrigation outdoor use (car washing, swimming pool) may account for the difference. If not, jiggle your irrigation estimates up or down by a "fudge factor" to match your actual use records. We'll revisit irrigation in Chapter 5.

A Note on Lawns

Unfortunately, turf accounts for the bulk of the irrigation need in the typical landscape, hence lawn greywatering is a popular violation of common sense greywater safety rules.

The only acceptable ways to irrigate lawns with greywater are via Automated Sand Filtration to Subsurface Emitters or Septic Tank to Subsurface Drip (Chapter 8)—well beyond what most folks are likely to install. We suggest replacing most of your turf with something else and replacing what's left with a water-conserving grass such as a Tall Fescue variety. Water this with the freshwater you save from using greywater elsewhere, or just let your lawn go dormant when there's not enough rain to sustain it. (See also Error: Surface Greywatering of Lawns, Chapter 11.)

Millionaire's greywatered lawn in a Los Angeles study. One side is irrigated with greywater through subsurface drip, the other with potable water. It's absolutely impossible to tell the difference. This system would be perfect for, say, a high school with dozens of showers and acres of turf.

Assess the Climate and Forces of Nature

Take stock of the climate parameters in the Site Assessment Form. Note any forces of nature such as flooding, drought, fire, and high wind, which may affect the design of the system or landscape.

Assess the Regulatory and Social Climate

As mentioned previously, it is much easier to make a practical greywater system if you aren't in a situation where impractical greywater systems are legally mandated. Will your neighbors let you get away with a system that doesn't have a permit?

Appraise the People Part of the System

People are a key part of any greywater system. Who is going to design, build, use, and maintain it? How interested and dedicated are they? If a dedicated person is responsible for maintenance, the system can be more demanding and delicate and still work. If there is no skilled person to build or maintain it, the system must be simple. If the system is public, public parts will have to be idiot-proof.

Cost/Benefit Analysis

Say your septic tank is failing and you can defer its replacement indefinitely by diverting all your greywater from it. Or, you know in your gut that it is wrong to send wastewater past your parched plants to pollute the river or ocean, while using electricity to pump freshwater out of a sensitive ecosystem miles away to slake their thirst, when you could water them with a simple, cheap greywater system instead.

In these cases, as in most, precision analysis is not needed. That's good, because accurate cost/benefit analysis is complicated. For starters, it requires accurate assessment of irrigation need in each season.

Simple retrofit systems will likely pay for themselves in water savings alone. This is almost assured if water costs are high, if the system is designed to maximize reuse efficiency, and if you do the work yourself.

In new construction, complex, contractor-installed systems can pay for themselves in water savings alone with as little as 100 gpd of greywater, if well-designed, and especially if *all* costs and savings are included. Often, economic structures preclude many costs and benefits from being felt. Individuals typically don't directly experience a cost savings from sending less water to the sewer, for example. Simplified cost/benefit analysis (see Table 2.7 for an example) does not consider lost interest, inflation, or detailed analysis of actual irrigation efficiency. If you have a compelling reason, you can take all these things and more into account.[13]

Revisit Your Goals

As you go through the design and installation of your greywater system, keep one eye on your goals. If new information or realizations outdate them, change them. For example, if you want a deeply ecological lifestyle but also want a perfect, classy system without suffering any cost or inconvenience, you may realize that this is self-contradictory. To make the transition from fantasy to reality, you'll have to let go of something. If you're working with a team, be sure to let the others know how the goals have changed.

TABLE 2.7: SAMPLE SIMPLIFIED COST/BENEFIT ANALYSIS
(FOR REAL WORLD EXAMPLE #2)

Savings	Value
Leachfield replacement deferred forever	**$5,000**
Longer interval between septic pumpings for 20 yrs	**$600**
Theoretical freshwater saved per year	40,000 gal *(150 m^3)*
× reality factor (rainy season, system inefficiencies)	50%
Cost per gallon (Santa Barbara average)	$0.005
Value of water saved per year	$100
Value of water saved 20 yrs	**$2,000**
Total savings per year	**$380**
Savings over 20 yr system life (not counting value of owner labor)	$6,790
Savings over 20 yr system life (counting cost of owner labor, or paying someone else to do it)	*$5,995*
Costs	
Cost of parts	$400
20 hrs laborer @ $8/hr	$160
5 hrs plumber (collection plumbing) @ $50/hr	*$250*
53 hrs owner labor inside & outside valued @ $15/hr	$795
Labor	$410
Total out of pocket	**$810**
Net savings	
Payback years (not counting owner labor)	**2.1**
Payback years *(counting cost of owner labor, or paying someone else to do it)*	**4.2**

By comparison to this Branched Drain, an Automated Sand Filtration to Subsurface Emitters system easily costs $5,000 in parts and labor. In addition to requiring substantial electricity and ongoing service, major parts require replacement after a few to several years (1998 costs).

Chapter 3: Design a Greywater System for Your Context

Now that you've loaded up your Site Assessment Form, site map, and/or brain with information, it is time to put it to work. If your brain is *over*loaded, consider skimming ahead to the next page, where the greywater basics track picks up again. In this chapter, you're going to consider how to integrate your greywater system with the landscape and other systems, address health considerations, and figure out how to best match your greywater sources with your irrigation or treatment/disposal area.

Integrate Greywater with Other Systems

This section only applies if you care about optimizing everything. In particular, it doesn't apply if you just want to dispose of the greywater.

Greywater systems, even more than other aspects of natural building, benefit from a systems approach to design. This is because they connect to so many other systems (Figure 3.1).

Sometimes when a greywater system is a retrofit, there isn't much you can do about

FIGURE 3.1: GREYWATER SYSTEM DESIGN COORDINATION

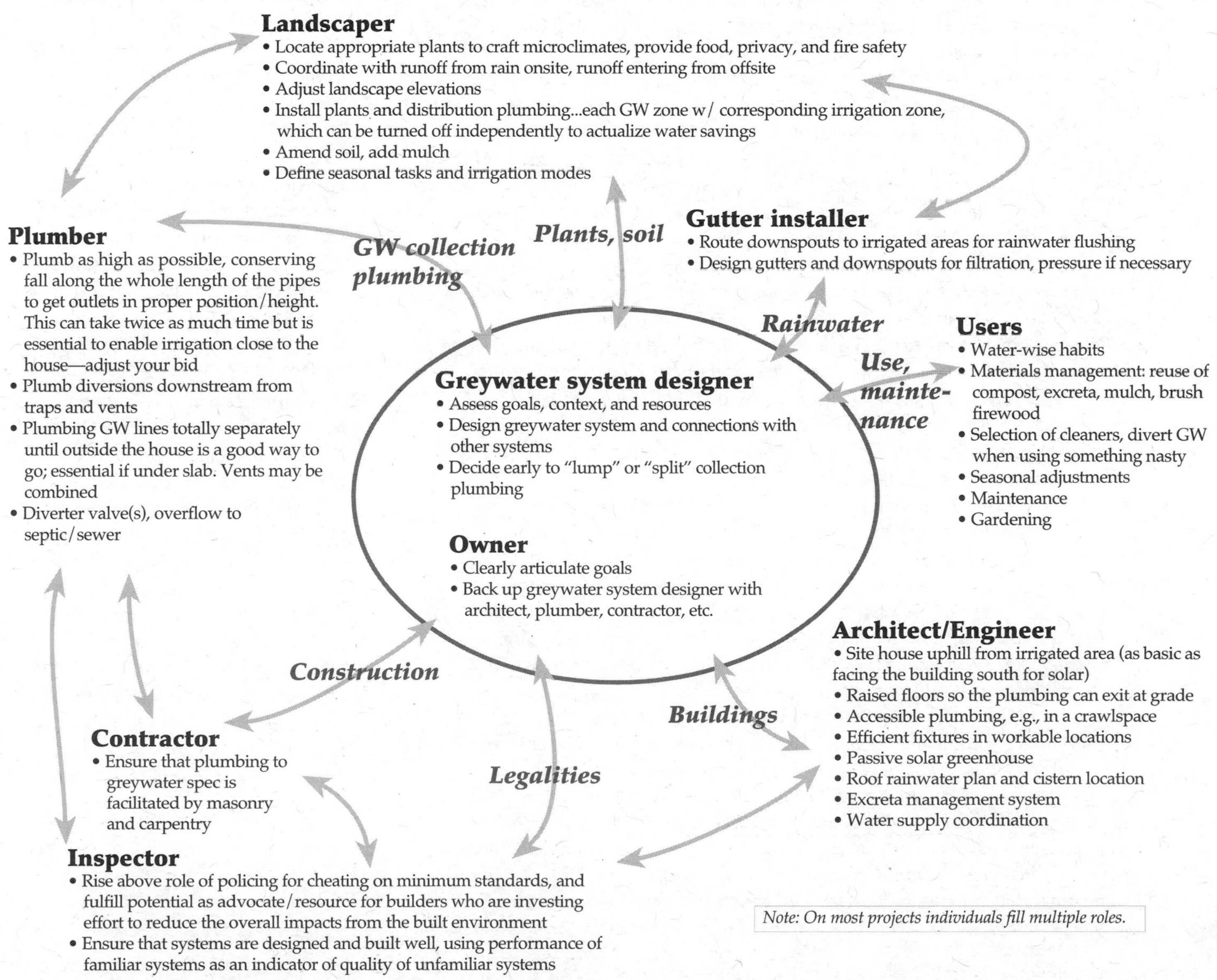

Landscaper
- Locate appropriate plants to craft microclimates, provide food, privacy, and fire safety
- Coordinate with runoff from rain onsite, runoff entering from offsite
- Adjust landscape elevations
- Install plants and distribution plumbing...each GW zone w/ corresponding irrigation zone, which can be turned off independently to actualize water savings
- Amend soil, add mulch
- Define seasonal tasks and irrigation modes

Plumber
- Plumb as high as possible, conserving fall along the whole length of the pipes to get outlets in proper position/height. This can take twice as much time but is essential to enable irrigation close to the house—adjust your bid
- Plumb diversions downstream from traps and vents
- Plumbing GW lines totally separately until outside the house is a good way to go; essential if under slab. Vents may be combined
- Diverter valve(s), overflow to septic/sewer

GW collection plumbing

Plants, soil

Gutter installer
- Route downspouts to irrigated areas for rainwater flushing
- Design gutters and downspouts for filtration, pressure if necessary

Rainwater

Users
- Water-wise habits
- Materials management: reuse of compost, excreta, mulch, brush firewood
- Selection of cleaners, divert GW when using something nasty
- Seasonal adjustments
- Maintenance
- Gardening

Use, maintenance

Greywater system designer
- Assess goals, context, and resources
- Design greywater system and connections with other systems
- Decide early to "lump" or "split" collection plumbing

Owner
- Clearly articulate goals
- Back up greywater system designer with architect, plumber, contractor, etc.

Architect/Engineer
- Site house uphill from irrigated area (as basic as facing the building south for solar)
- Raised floors so the plumbing can exit at grade
- Accessible plumbing, e.g., in a crawlspace
- Efficient fixtures in workable locations
- Passive solar greenhouse
- Roof rainwater plan and cistern location
- Excreta management system
- Water supply coordination

Construction

Buildings

Legalities

Contractor
- Ensure that plumbing to greywater spec is facilitated by masonry and carpentry

Inspector
- Rise above role of policing for cheating on minimum standards, and fulfill potential as advocate/resource for builders who are investing effort to reduce the overall impacts from the built environment
- Ensure that systems are designed and built well, using performance of familiar systems as an indicator of quality of unfamiliar systems

Note: On most projects individuals fill multiple roles.

Cont. from p. 9

Health Considerations

In practice, the health risk of greywater use has proven minimal. It is, after all, the water you just bathed in, or the residue from clothes you wore not long ago. Despite all sorts of grievous misuse (brought on in part by lack of useful regulatory guidance), there has not been a single documented case of greywater-transmitted illness in the US. Nonetheless, greywater may contain infectious organisms. It's poor form to construct pathways for infecting people, and totally unnecessary. Keep this in mind when designing and using a system.

All greywater safety guidelines stem from these two principles:

1. *Greywater must pass slowly through healthy topsoil for natural purification to occur.*
2. *Design your greywater system so no greywater-to-human contact occurs before purification.*

Here are examples of possible health-related greywater problems and their solutions:

- ***Direct contact or consumption**—Solutions: Carefully avoid cross-connections (accidental connections between freshwater and greywater plumbing). Label greywater plumbing, including greywater garden hoses. Use gloves when cleaning greywater filters. Wash your hands after contact with greywater.*
- ***Microorganisms on plants**—Direct application to foliage can leave untreated microorganisms on surfaces. Solution: Don't apply greywater to lawns, or to fruits and vegetables that are eaten raw (e.g., strawberries, lettuce, or carrots). Greywatering fruit trees is acceptable if the greywater is applied under mulch.*
- ***System overload**—Greywater systems are safest when reusing water that is fairly clean initially. Solutions: Greywater should not contain water used to launder soiled diapers or generated by anyone with an infectious disease. In both cases, greywater should be diverted to the septic tank or sewer. Also, don't store greywater; use it within 24 hours, before bacteria multiply. Finally, if you are having a party and 50 people are going to use a system designed for two, consider diverting greywater to the sewer for the night.*
- ***Breathing of microorganisms**—Droplets from sprinklers can evaporate to leave harmful microorganisms suspended in the air, where people may breathe them. Solution: Don't recycle greywater with sprinklers.*
- ***Chemical contamination**—Biological purification does not usually remove industrial toxins. Toxins will either be absorbed by plants or pollute groundwater. Many household cleaners are unsuitable for introduction into a biological system. Solutions: Don't buy products that you wouldn't want in your greywater system. Divert greywater that contains chemicals so they poison the sewer or septic instead.*
- ***Contamination of surface water**—Greywater needs to percolate through the soil, or else it might flow untreated into creeks or other waterways. Solutions: Discharge greywater underground or into a mulch-filled basin to contain it and slow its movement toward surface waters or groundwater. Don't apply greywater to saturated soils. Apply greywater intermittently so that it soaks in and the soil can aerate between waterings. In general, greywater that is confined subsurface or within mulch basins at least 50' (15 m) from a creek or lake is not a problem.*
- ***Contamination of groundwater or well**—It is all but impossible to contaminate groundwater with a greywater system, as the treatment capacity of the topsoil is so enormous. Over 90% of plant roots and beneficial microorganisms are in the top few feet of soil, above most septic leachfields. However, if you have a poorly sealed well, greywater running over the surface could potentially pour into it. Solution: The CPC/UPC greywater codes, which are just crude adaptations of septic tank codes, call for 100' of separation between location of greywater application and a well, same as a septic. Probably half of this is sufficient.*

Six Factors for Good Natural Purification of Water or Wastewater

Observe enough engineered and natural water systems...and you realize we'd best do it as much like nature as possible. Here are some principles to guide ecological water treatment:

1. ***Plenty of contact time**—The longer the water is in contact with bacteria and plant roots, the better. To increase contact time, reduce flow and/or increase area.*
2. ***Plenty of microsurface**—The more microsurface with beneficial bacteria growing on it, and the more plant roots, the better. Loamy soil has thousands of times more surface area than gravel.*
3. ***Moisture, oxygen, and nutrient levels which support growth/survival of roots and bacteria**—If the system is totally dry long enough for the bacteria and roots to die, then gets spike loading, then is dry again, the treatment won't be as good. If the soil is saturated (no air) for more than 24 hours, the dissolved oxygen will be consumed.*
4. ***Apply wastewater as close to the surface as possible, without causing an unsanitary condition**—The top of the soil has a purification capacity thousands of times greater than 3' down, because 90% of the life is within a foot of the surface.*
5. ***Appropriate plants**—You don't have to worry about the bacteria—if the conditions are right, one will turn into trillions. However, it is generally helpful to actively manage plants to ensure there are the right number of the right kind. Evergreen plants are active all year. If you expect constant, water saturated conditions, use wetland plants, which pump oxygen out their roots.*
6. ***Warmth**—The warmer it is, the better the treatment. The rule of thumb is that for each additional 10°C you get twice the treatment.*

how it "fits in." In other cases—such as new construction on raw land—you can design a greywater system together with all these other systems. When you do this, you gain *economies of wholeness*. For example, a house can catch rainwater from its roof, heat it mostly with the sun, provide supplemental heat from burning yard trimmings, use the water in super-efficient fixtures, then irrigate trees with this water. These trees, if carefully selected and positioned, purify the water before it recharges the groundwater, and also provide shade from heat, shelter from cold wind, privacy from neighbors, and delicious fruit.

Coordinate with Others

If a team is to install the greywater and related systems, and you want an optimal outcome, it will be necessary to coordinate everyone's efforts.

If you're doing all the work yourself, you'll still need to coordinate, but it will be easier with all the specialists under your one hat.

General Landscape Design Points

Here's the short version of my checklist (there's more on these points in Chapter 5):

- **100% rainwater and runoff infiltration**—Form landscapes with basins, swales, mulch, and such rich, loamy soil that *all the rain that falls on the property soaks in*. Or more—landscapes can absorb "runon" from surrounding surfaces. This yields benefits including flood control, reduced irrigation need, and improved quality and quantity of natural surface waters. (Our book *Water Storage* covers storage of water in soil, groundwater, and seasonal ponds.)
- **Rainwater/runoff flushing**—Particularly in places with clay soil or low rainfall, directing rainwater into the system in a controlled way flushes harmful salts out of the soil.
- **Irrigation near the house**—In most cases, though you don't want to wet your actual foundation, you want the greywater to irrigate plants right around your home, so you can live in the oasis thus created, not way uphill from it. This usually means *watching elevations really carefully*.
- **Microclimate modification, outdoor living spaces**—Plants can dramatically improve the climate, indoors and out. I design landscapes with an outdoor living space that's sunny and warm and another that's shady and cool, both easily accessible from the house. Identify places that naturally have these attributes, then accentuate them by shading or reflecting sun, screening wind, and irrigating. (Evaporation provides outdoor air conditioning.)
- **Edible vs inedible landscaping**—My philosophy is to only water things I can eat—and a few things of overpowering beauty. If a non-edible plant is required, I pick one that will grow without irrigation.
- **Privacy plantings**—If you like privacy, hedges are a green, economical, and beautiful way to achieve it. Greywatering hedges makes sense, especially when space is tight or later construction might tear up a system in the middle of the property.

Connect Greywater Sources with Irrigation or Treatment Area

On your site map, review the gallons of greywater available per week from each group of fixtures, and where their drain plumbing goes.

Then look at the areas you've shaded on this map to indicate where irrigation is needed, or where the greywater could go for treatment/disposal.

Now consider which greywater sources could be routed to which areas. For irrigating, you're generally looking for plants that are close to the house and, with a gravity system, reachable by a continuously downhill pipe run.

Lump or Separate the Greywater Flow?

This material applies to everyone.

The first design decision is whether to go with one combined line, or separate the flows first by not combining them. This decision determines the layout of the collection plumbing in the house and, once done, is not easy to change. How do you decide?

Separate Flows

If you want a very simple system and/or don't care about reuse efficiency, separate flows are for you.

The benefits of not combining flows are: simpler plumbing, less fall required (because the pipes don't have to go so far, first to join all together, then back to their own plants), and the ability to take advantage of the unique flow characteristics of individual fixture sets. Examples: An upstairs bathroom might be able to water a part of the garden that is too high to reach otherwise. If you tried to get the greywater up there after the flows were combined, it would bubble up the downstairs shower drain instead. Or, a washing machine can supply a higher part of the landscape by using its own pump to raise the water.

Separate flows reliably ensure some degree of irrigation reuse. No matter what, the water comes out in multiple outlets—for example, one spot watered by dishwashing, another by clothes washing, yet another by bathing.

Separate flows may be preferable to combined flows if:

- **You don't care about reuse efficiency.**
- **The house is large and sprawling.**
- **Rerouting plumbing from one side of the house to the other is difficult or impossible**—E.g., with slab on grade construction.
- **The land around the house is very flat**—There may not be enough fall to join flows and then send them back to plants.
- **The house is more than one story and portions of the yard are above the ground floor**—If the greywater from the upper floor(s) is plumbed separately, it can gravity flow to more of the yard.
- **You have very little fall between sources and plants**—Separate flows may be your only option.

(Figures 9.1, 9.4a, 9.4d, 12.1, and 12.3 show systems with separate flows.)

Combined Flows

If you care about reuse efficiency, combined flows are more likely your route.

The advantage of combining greywater flows in the collection plumbing before splitting is that variations in flow from different fixtures average out. This greatly simplifies supplemental freshwater irrigation. This is especially true if some fixture sets have highly variable use, for example a guest bathroom.

Combined flows work out better if:

- **Your goal is high reuse efficiency for irrigation**—Especially if your plants will rely exclusively on greywater.
- **Your system requires filtration or pumping**—To avoid the need for multiple pumps and filters.
- **You want to be able to switch between multiple irrigation zones**—See below.

(Figures 9.4b and 9.4c show examples of flows combined before being split.)

Sketch your lumped and separated plumbing alternatives on your site map, and see how they look. Separated flows will generally look easier on paper and be easier to build. The *management* is trickier, as described above. Once the combined or separate decision is made, you can design the collection plumbing.

Multiple Greywater Zones

Greywater comes out many times a day, as it is generated. The only drying out periods are when everyone is asleep or on vacation—and not even then if there is a leak. Having more than one irrigation zone, and the ability to alternate greywater between zones with valves or to reroute greywater into the sewer or septic system, is advantageous in some situations, to:

- maintain aerobic conditions in irrigation zones
- divert greywater to disposal when irrigation is not needed
- divert greywater to disposal, or spread it more widely, when a steep slope could get oversaturated and slide
- fit available greywater more closely to irrigation needs

In one system (see Real World Example #1), the homeowners observed that their citrus trees did well with continued watering for 2–3 weeks without a drying out period. However, their stone fruit trees seemed stressed by the long irrigation periods.

Particularly with *low perk soil* and/or sensitive plants, it would be good to make provision for drying out periods between watering. Designing the greywater system to provide only a portion of the plants' water needs helps solve this problem (see Table 5.1). Another solution is to switch from one zone to another with a diverter valve or valves. Also, greywater can water two zones when it is sufficient to meet all plants' needs, then only water one or the other when it can't meet the needs in both. The obvious problem is remembering to switch the valve. An automatic controller is great for sending a few days' greywater to one zone, then another. These cost a few hundred dollars. Unless the greywater is pressurized, the valves themselves cost hundred of dollars each, as well.

Provide for Maintenance and Troubleshooting

I'm a believer in making every part of the system find-able, accessible, maintainable, replaceable, and flexible. Specifics on this important point will be discussed throughout.

Practice Ecological Design

Here's a general recipe for the practice of integrated, ecological systems design:

- **Get a global, yet detailed view of your context**—Ecological design is based on a view of the world that is wide angle, with high resolution zoom and x-ray vision to relevant details, clear as to scale, rooted in the past, and fast fowardable into the future.
- **Get clear on goals.**
- **Inventory issues.**
- **Inventory resources.**
- **Creatively use resources to resolve issues.**
- **Have fun doing it...**

For much more on this topic, see *Principles of Ecological Design*.[15]

Relate Well with the Natural Water Cycle

Here's a very brief overview of principles for plumbing your home properly into the global nutrient and water cycles:

- **Leave as much of the work as possible to nature**—The more humans intervene, the more likely the overall system will get thrown out of whack.
- **Work on solving several problems with one design**—Include comprehensive as well as specialist perspectives. More can be gained from improving connections between systems than from improving systems.
- **Divert a small amount of water.**
- **Divert just after natural purification**—If water from springs, rainwater harvesting, or wells is diverted after natural purification and before it is contaminated, little or no additional treatment is needed.
- **Divert from an elevation above the use point**—Or as little below it as possible, so less energy is needed for pumping. You can also save energy by using ultra low pressure plumbing.
- **Use water efficiently**—REDUCE comes before reuse in the hierarchy of ecological materials management. Always consider efficient fixtures before looking to reuse water from them.
- **Cascade**—Sequence uses so water cascades from those uses which require the cleanest water to those which tolerate the dirtiest.
- **Use super-efficient fixtures**—E.g., wood burning bathtub, eco-luxury bathing chamber.
- **Add used water and nutrients back into the water cycle at large just before natural purification**—Greywater systems, composting toilets, Green Septic systems, compost, mulch...
- **Absorb all runoff**—Permeable surfaces, vegetation cover, mulch, basins, and swales.
- **Rigorously confine incompatible materials (motor oil, solvents) to their own industrial cycles**—Add to water only a moderate quantity of substances which biodegrade into plant nutrients or nontoxins and nothing else.

Chapter 4: Greywater Collection Plumbing

If you are only going after the washing machine water, you can skip down to Collecting Pressurized Greywater, and skip the plumber as well. If you are building a very simple, unpermitted dwelling, and want to save as much money and resources as possible, I suggest you read this whole chapter to sharpen your plumbing understanding, then go with Radical Plumbing (described at the end).

Collection plumbing takes the greywater from various generation points in the house to a point or points just outside. From there, distribution plumbing takes the greywater to plants. For gravity collection plumbing, codes provide useful guidance, and the services of a plumber are helpful. (For distribution plumbing in the landscape, CPC/UPC-style codes and a plumber are likely to be a hindrance.)

Greywater collection plumbing is largely independent of the type of system used to distribute greywater. However, you should select and design your distribution system before you start sawing under your house. Collection plumbing does vary a great deal depending on whether it is for gravity sources, appliances with their own pump (washer or dishwasher), or a greywater-only (Radical Plumbing) system. This chapter presents general plumbing principles and then looks at each of these three contexts.

Surge tanks, filters, pumps, and disinfection (if any) go between the collection and distribution plumbing.

General Greywater Plumbing Principles

For all greywater users, it is helpful to understand some plumbing principles, 1) so you understand how greywater seemingly violates the laws of physics that govern freshwater flow, and 2) so you can communicate unique greywater considerations to your plumber.

There isn't a well developed code for some of the strange things greywater plumbers have to do, so understanding general principles[14] will help navigate this *terra incognita*.

When to Get Professional Help

Help from a plumber is recommended for collection plumbing.

It is said that "plumbers safeguard the health of the nation." This is not an exaggeration. Don't let the senselessness of some greywater laws delude you into thinking that *all* plumbing codes are for weenies. The place where codes lose credibility is in the garden, where greywater can be treated better and more ecologically than anywhere else. Under the house, use a plumber and deviate from code only when you know what you're doing, or you may be sorry.

Squander No Fall

Fall is the vertical distance between greywater source and destination. Every trap, pipe run, joint, valve, etc. uses up some fall in order to keep the water flowing downhill. Proper slope is the #1 driving factor behind collection plumbing design and installation (and Branched Drain distribution plumbing as well). For 90% of greywater installations, fall is a **critical, system-limiting factor to be conserved religiously**.

Tests with clear pipe have shown that unpressurized, crud-laden water flows best in pipes that are either vertical or that slope ¼" per foot (2%). At this slope the solids and the water move at the same speed. With shallower slope, water won't flow well. At steeper slope, especially over long pipe runs, the water tends to get ahead of the solids and leave them behind. If the next water down the pipe doesn't scour them off, the solids build up and clog the pipe. However, field accounts of this problem are rare with greywater systems. Most people get away with a bit more slope, and even (horrors!) a bit less, if the length is short. Be sure there is cleanout access nearby. Going uphill

Collection Plumbing and Inspections

Use a skilled plumber for the house plumbing when the system is to be inspected. The typical inspector's eyes glaze over in the yard and they don't say much, perhaps not wishing to reveal ignorance. Then, they hold your feet to the fire on the inside plumbing where they really know how to apply the code, zapping you on the slightest deviation.

Warn your plumber and use this phenomenon to your advantage. Time and again it has been noted that if the standard plumbing is impeccable, inspectors assume the novel part is done to the same high standard and wave it through. Conversely, if the collection plumbing is flaky, they'll assume you did the distribution plumbing wrong, too.

is out of the question, even for a short distance. Crud will surely settle in the bottom of any U-shaped section and block the flow (except in traps, where the vertical drop just before the trap blasts the crud through; see Figure 4.1).

You've got to really watch your plumber because it is extra work to fanatically conserve fall, and they will probably be reluctant to do it.

Fall conservation may be a point of contention with your plumber as it can quadruple the design and installation effort. Septic tanks and sewer lines are typically *much* deeper underground than greywater lines, and their destination is unlikely to change. Therefore, plumbers are accustomed to conserving no more fall than is necessary to make it to the sewer/septic line, which is not much. Extra fall conservation tends to be viewed as a major waste of effort.

Greywater collection plumbing differs from conventional septic/sewer collection plumbing in two important respects. First, the destinations are generally much higher. Second, it is highly likely that as plants grow and die and new ones are added over the long life of the collection plumbing, you will want to send the greywater somewhere else at a slightly higher elevation.

Achieving proper slope often involves considerable heroics. I often chip through concrete, replumb far upstream, or do whatever else is necessary to get the right slope. I had to replumb the entire underneath of the house in Real World Example #1 to gain half a foot of elevation in the greywater pipes. Systems that require a surge tank require more fall.

Explain to your plumber clearly what you need and the logic behind it. Ask that they make every connection and pipe run as high as possible.

Build for Future Flexibility

This is important for all systems.

At every opportunity, leave flexibility to reconfigure the system in the future. Where possible, leave *enough pipe* between fittings so they can be sawn apart and reconfigured without throwing the whole assembly away. This may conflict with conserving fall at times, but do it whenever you can afford the fall. Your plumber will understand this issue. Expensive, complicated components such as valves and dipper boxes should be installed with no-hub connectors (removable couplings), or silicone sealer instead of ABS glue on street fittings. (Silicone should be allowed to dry 24 hours before use.) Outdoors, they can be friction-fit connections (that is, without glue). This enables these valuable components to be easily serviced, reused, reconfigured, or replaced.

Divert Greywater Downstream from Traps and Vents

This issue is specific to greywater collection plumbing. Once in the yard, traps are not needed, and venting is usually not an issue (any necessary venting is usually accomplished back through the house plumbing).

Vents and traps are unlikely to be a point of contention with your plumber. Running greywater separately from toilet water is unusual, but plumbers can do a great job on traps and vents with their existing knowledge base.

One of the features that distinguish modern plumbing from earlier, cholera-epidemic-era efforts is the vented p-trap, so named because it looks like a P on its side. The p-trap, by virtue of the small pocket of water it always holds, blocks all of the noxious gases and pernicious filth in the sewer/septic from escaping up through the pipe into the house (Figure 4.1).

Careful distance, elevation, and size relationships must be maintained to prevent unforeseen p-trap failure. For instance, the p-trap must be vented just downstream to break the vacuum created behind a flow of water that would otherwise siphon the little pocket of water right out of it. (The s-traps in some older homes are unvented and can suffer this problem, which can be alleviated by making the s-trap a larger size pipe.) The vent must be within the critical distance, so water filling the drainpipe won't block air's path to the vent.

FIGURE 4.1: P-TRAP AND CRITICAL DISTANCE

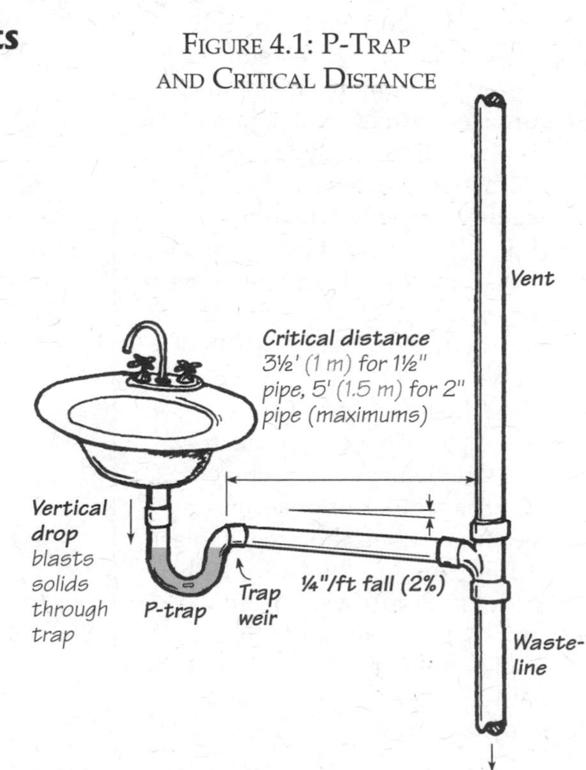

Venting for greywater collection can be totally separate from the toilet vent. Or, code allows you to tie into the toilet vent 12" *(30 cm)* above the spill point of the highest fixture served by the vent (see Figure 4.3). This eliminates the possibility of blackwater backing up through the common vent and into the greywater system. (In some areas you may be allowed to use auto vents, a delicate kind of vent with a 1-way valve that lets air in but not out. These can go inside the house.)

Traps and vents are unnecessary if you cut all ties to the septic/sewer and do Radical Plumbing.

FIGURE 4.2: A CLEANOUT

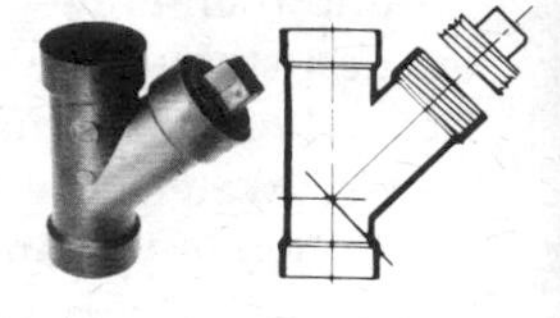

Provide Cleanouts and Inspection Access

This is important for all systems.

Keep ease of service in mind. You don't want to build *anything* that cannot be inspected or serviced—especially plumbing that could last the life of your house.

Code calls for a cleanout, a place you can run a plumber's snake into the pipe to clear it, every 135° of cumulative bend (a 90° plus a 45°, for example). Snaking down the pipe (with the flow) is preferable to snaking uphill. However, either direction works. Plumbers incorporate cleanouts in the collection plumbing as a matter of course.

The face of some 3-way valves is removable,[s2] providing cleanout access in all three directions. Care must be taken to not wreck the inside sealing surfaces of the valve with the snake. A dipper distribution box (p. 82) provides excellent access to every pipe connected to it, without worry of wrecking a seal.

A cleanout right after the line exits the house is a convenient feature. If possible, locate this by a rain downspout and you have a perfect setup for periodic flushing of salts as well as snaking. Don't hook the downspout into it permanently. That is illegal, and will almost certainly overload the system at some point.

Design for Easy Maintenance and Troubleshooting

This is important for all systems.

A well-designed system provides auditory and visual signs of its distress. Slow drains or water running loose in the yard are not catastrophic failures. Water overflowing into the house, greywater siphoning into the potable water supply, or a $500 pump burning out *are* disasters. An inaudible pump is much more likely to burn out when it strains, unheard, against a clogged distribution line. A system that overflows silently and cleanly into the sewer when the filter clogs is not the most efficient. Weeks' worth of greywater may be wasted each time the filter clogs, before it occurs to anyone to check it.

Greywater systems may require regular filter cleaning or other troubleshooting. Table 4.1 is geared to the Drum with Effluent Pump system, but the principle of thinking "what will happen if..." is valid for all system types. Design your system so it provides you with the

TABLE 4.1: FAILURE MODES FOR FILTER AND PUMP SYSTEMS

Failure mode	Design feature	What will happen
Primary filter clogs (before surge tank; occurs normally with use)	No overflow (preferred)	Drains will run slowly, then not at all, letting you know it's time to clean the filter. Disgusting water may back up into lower fixtures such as baths and showers.
	Overflow	Water will flow out overflow; *if* you can hear it, you'll know it's time to clean the filter.
Secondary filter clogs (on the way out of surge tank)	No pressure relief	Pump will work against clogged filter until you notice or it dies. Note: Most submersible pumps can operate against a blocked line for a long time without damage, as long as they are submerged.
	Pressure relief (preferred)	Water will be pumped out of pressure relief until you unclog filter.
Pump failure	N/A	Same as clogged primary filter, above
Distribution lines or emitters blocked	N/A	Same as clogged secondary filter, above
Check valve held open by hair	N/A	Water may leak downhill from distribution lines back into surge tank, causing pump to cycle on over and over.
Surge capacity exceeded before filter	No collection line overflow	Lower elevation fixtures overflow.
Surge capacity exceeded in tank	No surge tank overflow	Lower elevation fixtures overflow.
Irrigation area saturated	No disposal alternative	Possibility of waterlogging, greywater runoff

proper cues—noticeable, but not catastrophic—when it needs attention. Incorporate an easily switchable bypass to the septic/sewer for when your system needs maintenance or you don't want to use greywater. In addition to a manual bypass, you want an overflow that splashes conspicuously but harmlessly outside if the system is non-operational (a more health-department-friendly option is an alarm that sounds when water overflows to the regular drain).

Collecting Pressurized Greywater

You can skip this if you are not diverting laundry or dishwasher water in its own separate system—pressurized greywater sources can also be vented of pressure through an air gap and treated as gravity driven sources.

For pressurized greywater sources (washing machines, automatic dishwashers), collection plumbing can be as simple as shoving the outlet hose through a window or a hole drilled in the wall to a surge tank or distribution plumbing just outside (see Laundry Drum, Chapter 7). In these scenarios, there is an air gap at the end of the appliance drain hose, as with conventional plumbing. The advantage is that the load on the pump is the same minimal load the appliance was designed for.

Appliance pumps can also be used to move greywater long distances horizontally and short distances vertically. The appeal is that you don't have to buy any additional hardware, surge tank, or filter. (In some instances, however, this can result in premature pump failure.) For details, see Laundry to Landscape, Chapter 7.

Collecting Gravity Flow Greywater

Read this closely if you are cutting into your plumbing or building a new house.

To collect water from gravity driven greywater sources such as sinks, tub, and shower, major replumbing is necessary.

Greywater collection plumbing may be fairly straightforward if drainpipes are easily accessed, or impossible if they are entombed under a concrete slab foundation. In any case, using a **professional plumber to plan and execute collection plumbing is highly recommended.** Without the correct fall, venting, and traps, the system may not drain satisfactorily or even be safe. In some areas, it is illegal to modify your own plumbing for these reasons.*

Collection plumbing is a good place to stick to code. An intelligent plumber unfamiliar with greywater can successfully ensure that the collection plumbing's fitting-by-fitting detail is done correctly. However, for the vast majority of plumbers, it is *critical* to guide them on the overall design and check frequently to be sure that they do a good job at the points where greywater plumbing diverges from conventional practice. Otherwise you will probably have to redo something. If you do want to tackle this yourself, refer to a plumbing book and the distribution plumbing tips in Chapter 10, and/or have a plumber check your work.

For gravity sources, the plumbing that routes water from the fixtures to the outside will be conventional except for these few respects:

- **Rigorous conservation of fall**—This impacts the design of every joint and pipe run. You must hover over your plumber to ensure enough height where pipes exit the house to irrigate intensive plantings right around the house.
- **Toilet water will be separate**—For new construction, I recommend *running greywater in separate pipes to the outside* even if you join them there. This gives you flexibility to add a greywater system in the future.
- **Diverter valves**—Installed to enable greywater to be diverted to septic or sewer in case the soil is waterlogged, the system is down, or you want to use some nasty cleaners that you don't want on your plants.

You may wish to skip the next two pages unless you are a builder or unfazed by technical stuff.

Still reading? These pages give a few examples of how collection plumbing can be done. Every installation is different. There is more on collection plumbing for new construction and slabs in our Builder's Greywater Guide.[6]

Continue p. 41

The exact rules vary by state/county laws. In general, you, the homeowner, can work on your own plumbing, but you can't hire yourself out to do plumbing, and in many places you can't hire a non-plumber (i.e., neighbor's kid, migrant laborer) to do your plumbing.

FIGURE 4.3: HOW TO TIE INTO PLUMBING

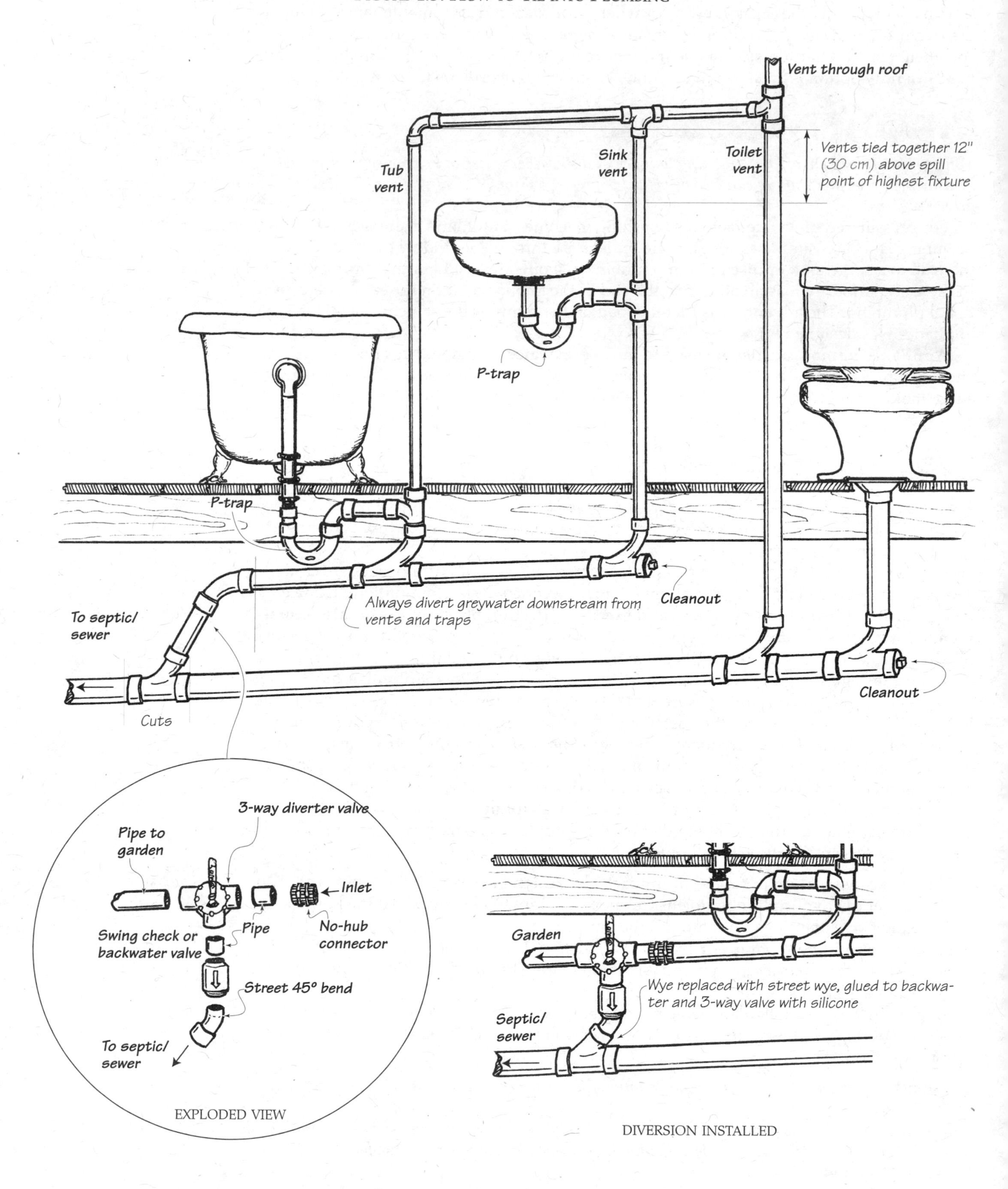
Vent through roof
Tub vent
Sink vent
Toilet vent
Vents tied together 12" (30 cm) above spill point of highest fixture
P-trap
P-trap
Cleanout
Always divert greywater downstream from vents and traps
To septic/ sewer
Cleanout
Cuts
3-way diverter valve
Pipe to garden
Inlet
Swing check or backwater valve
Pipe
No-hub connector
Street 45° bend
To septic/ sewer
EXPLODED VIEW
Garden
Wye replaced with street wye, glued to backwater and 3-way valve with silicone
Septic/ sewer
DIVERSION INSTALLED

FIGURE 4.4: GREYWATER DIVERSION FROM CAST IRON DRAIN LINES

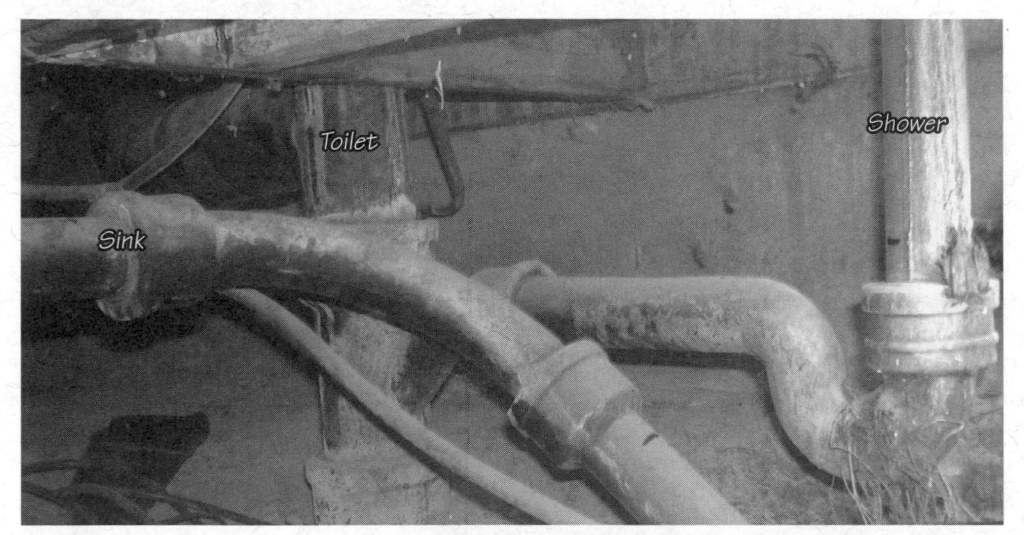

Before: *cast iron drain pipe for a bathroom in an old adobe home in New Mexico.*

Cutting the old iron pipe with a sawsall.

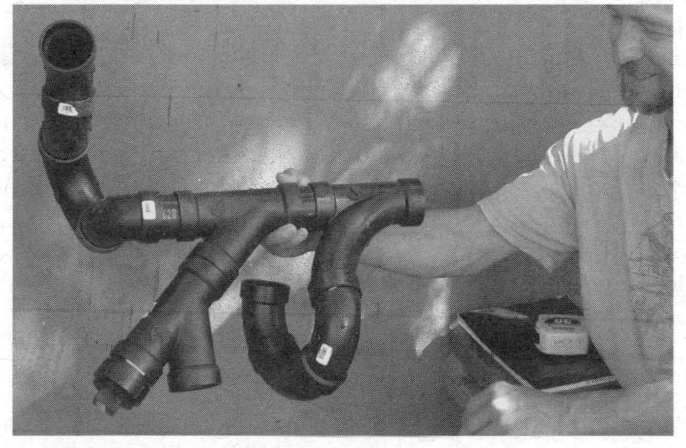

Greywater plumbing subassembled *(no glue).*

Greywater collection plumbing installed.

Surge Capacity

All systems need this.

Surge capacity, or the ability to handle peak flows, should be provided in every system. Systems without any flow bottlenecks (e.g., Branched Drain) have all their surge capacity in the landscape itself, while most other systems require tanks upstream of bottlenecks. We'll now look at all the sizes of surges (Table 4.2), and places surges can be accommodated: collection pipes, surge tanks, distribution pipes, and landscape.

TABLE 4.2: GREYWATER SURGE VOLUMES

Water source	Typical surge	gallons	cubic feet	liters
Washing machine	1 min each cycle	15	2.0	*57*
Tub draining	3 min	45	6.0	*170*
Low flow shower	10 min	20	2.7	*76*
High flow shower	10 min	50	6.7	*190*
Bathroom sink (low flow)	1 min	1.5	0.2	*5.7*
Kitchen sink (draining)	1 min	4	0.5	*15*

A bathtub can drain at 15 gpm, far in excess of what a pump or filter can accommodate. If the washing machine cycles on at the same time the bathtub is draining, 30 gal of water could flow out in 1 minute, followed by another 15 the next.[m] For most homes, 45 gal of surge capacity is sufficient. If you habitually launder and run the dishwasher while you are in the shower and your kids are in the tub, you need more. The overflow consequences of exceeding surge capacity have a major bearing on this decision. Systems that require a surge tank require more fall.

Pre-Filter Surge Capacity in Collection Plumbing

This only matters if you have a filter.

Surge capacity before a filter is given little thought in most designs and amounts to whatever space there is in the collection lines (see Table 4.3). Collection plumbing is usually sized to accommodate the flow if everything was draining at once. It is a rare greywater system that requires bigger than 2" pipe to achieve this. The volume of pre-filter surge capacity determines how soon you notice slow draining and need to clean your filter. A common amount is 10 gal *(40 L)*; if you have short collection lines and fall to spare, you can install a pre-filter surge tank. Design it so solids are flushed through, otherwise it will need periodic cleaning. Bigger pipe could accommodate surges but may not purge solids afterwards.

Surge Tanks

Typically surge tanks are located downstream of pressurized greywater sources and upstream of pumps, filters, and small distribution orifices that slow the flow: in other words, anywhere surge capacity is needed to prevent a problematic backup from a flow bottleneck.

The simplest surge tank is a 30–55 gal plastic drum (*100–200 L*—see photo under Laundry Drum, Chapter 7). Greywater corrodes metal drums after a few years. Empty plastic drums are UPS-able, increasing your scrounging range. Various industries generate surplus drums that they give away or sell cheaply. Reject any that held something toxic. Nurseries and hardware stores sometimes sell plastic drums. Many people use inexpensive plastic trash cans as surge tanks. But they soon become trash themselves when sunlight weakens them and the water weight splits them open. Drums with tight-fitting, large lids are ideal, but are much rarer than drums with two small bungholes. The latter work if you don't need to put a pump inside, or you can saw the whole top off, then rig it as a lid. Laundry sinks next to washers can be used as surge tanks.

Surge tanks that are buried during drought can suddenly pop to the surface when rains finally saturate the soil. If you bury a tank, anchor it well against floating. The upward force on an empty drum surrounded by water is considerable: almost 500 lbs *(200 kg)* for a 55 gal *(200 L)* drum. Certain soils also progressively cave in buried drums as the soil expands and contracts. Tanks designed to be buried don't have this problem.[s5]

Surge tanks often stink. Smell can be reduced by minimizing the holding time. Sealing the tank and venting it through the house's roof can minimize its noticeability. Storing greywater isn't a good idea; we cover the gory details in Error: Storage of Greywater, Chapter 11.

---

[m] ***Metric:*** *Bathtub can drain at 55 lpm. If washing machine cycles on while tub is draining, 115 L can flow out in 1 min, and 55 L the next min. A 170 L surge capacity is enough for most homes.*

Surge Capacity in Distribution Plumbing

Surge capacity in distribution plumbing is rarely a limiting factor. It does, however, affect evenness of distribution. Suppose you have 62 gal of surge capacity in 100' of 4" leachline and your pump cycles on with 8" of water in the tank and off at 3".[m] The few gallons it pumps out with each cycle will only wet the first few feet of the line. If it were 100' of ½" dripline, the first gallon would fill and pressurize it, and the rest would be pumped out evenly over emitters along the whole line. (This better distribution makes more work for the pump.)

To water with large, infrequent doses, improve distribution, and give the soil resting time, set the float switch so the pump cycles on when the surge tank is nearly full, and shuts off when it is as empty as possible. *Note: Greywater can go septic if it sits in a dosing surge tank while you're on vacation. Also, long float-switch tethers can tangle or hang up.*

Surge Capacity in the Receiving Landscape

The simplest place to accommodate greywater surges is in the landscape, as Branched Drain, Laundry to Landscape, and Landscape Direct systems do. For all systems, the surge capacity in the landscape—basins, swales, and soil—should be adequate to receive the largest expected peak flow (though some automated systems can be set to send excess greywater to septic/sewer). A few gallons of greywater landing on hard soil could run off into the street. But even 100 gal of greywater may not fill a deep mulch basin around a large tree to the point of surfacing or overflowing, even if the soil is full of rainwater.

As previously mentioned, the possible consequences of exceeding surge capacity affect system design. For example, if the whole area is mulched, any water that overflows a basin is still covered by mulch, so it's no big deal. For most systems, making the basins twice the volume of the expected surge is sufficient. For big systems, or stringently regulated systems, more sophisticated calculation of landscape capacity is described in Greywater Calculations in our *Builder's Greywater Guide.*

Dosing Siphon

A dosing siphon (Figure 4.5) turns many small doses of greywater into one large one, without a pump. If you try one, make sure the outlet line is small enough for an average surge to activate the siphon and not just dribble over the spill point; 1½" pipe should work. Accumulated sludge and scum could eventually block the outlet.

Caution: It is tricky to get homemade dosing siphons to work reliably.

Figure 4.5: Do-It-Yourself Dosing Siphon

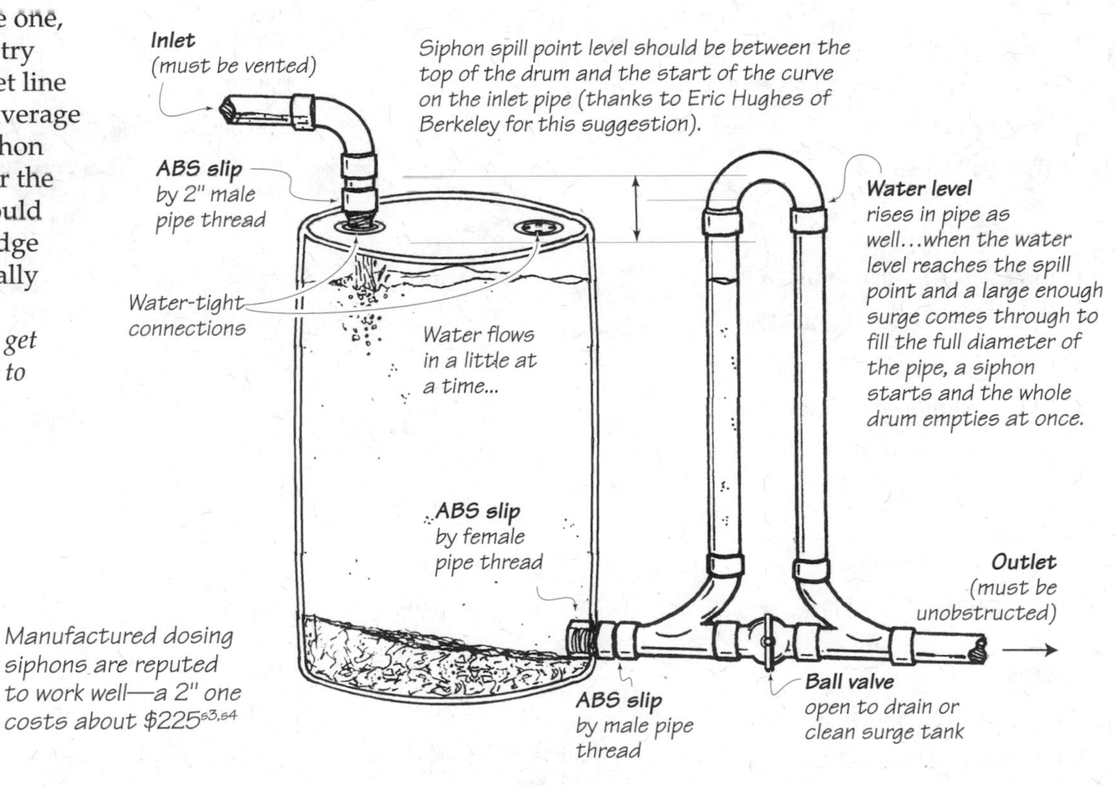
Inlet
(must be vented)
Siphon spill point level should be between the top of the drum and the start of the curve on the inlet pipe (thanks to Eric Hughes of Berkeley for this suggestion).
ABS slip by 2" male pipe thread
Water level rises in pipe as well...when the water level reaches the spill point and a large enough surge comes through to fill the full diameter of the pipe, a siphon starts and the whole drum empties at once.
Water-tight connections
Water flows in a little at a time...
ABS slip by female pipe thread
Outlet (must be unobstructed)
Manufactured dosing siphons are reputed to work well—a 2" one costs about $225[s3,s4]
ABS slip by male pipe thread
Ball valve open to drain or clean surge tank

[m]*Metric:* 230 L surge capacity in 30 m of 100 mm leachline, pump cycles on at 20 cm and off at 8 cm.

TABLE 4.3: GREYWATER PLUMBING PARTS

Environmental impact (bigger dot = more impact)	Parts	Greywater Application: Gravity collection	Greywater Application: Pressure collection	Greywater Application: Filtered distribution	Greywater Application: Unfiltered distribution	Sold by	$/ft or $/each (2008)	Flow rate gpm	Surge capacity gal/ft or gal/each	*L/m or L/each*	Normally used for; available from
	Rigid pipe										
•	¾" schedule 40 PVC			x		10, 20', *3, 6m*	$0.33	12	0.025	*0.31*	Drain Waste Vent (DWV); hardware store
•	1" schedule 40 PVC		x	x	x	10, 20', *3, 6m*	$0.49	15	0.041	*0.50*	
•	1½" schedule 40 ABS	x			x	20', *6m*	$1.09		0.100	*1.24*	
•	2" schedule 40 ABS	x			x	20', *6m*	$1.34		0.164	*2.04*	
	Flexible tubing										
•	½" HDPE dripline			x		100, 500'	$0.12	4	0.013	*0.16*	Drip irrigation; hardware store
•	¾" HDPE dripline			x		100, 500'	$0.29	8	0.022	*0.27*	
•	1" HDPE dripline			x	x	foot, 600'	$0.49	12	0.040	*0.50*	Potable water, landscape; hardware store
●	1" spa-flex[s6,s7]		x		x	foot, 50'	$1.18	15	0.041	*0.51*	Hot tubs; pool supply store[s6,s7]
•	¾" garden hose			x	x	50, 75'	$0.95		0.025	*0.31*	Landscape; hardware store
	Emitters										
•	Underground dripline			x		187.5'	$0.64	1.15	0.013	*0.16*	Septic leachfield; plumbing supply
●	Standard recycled HDPE infiltrator w/2 end plates			x		each (6' long)	$80.28		76	*287*	
	Valves										
●	1½" Jandy 3-way diverter valve	x	x	x	x	each	$59.00				Pools; pool supply,[s7] oasisdesign.net[s9]
●	2" Jandy 3-way diverter valve[s2, s9]	x				each	$59.00				
●	Electric switch for 2" Jandy valve	x				each	$250.00				
•	½" drip ball valve			x		each	$2.95				Drip irrigation; plumbing supply
•	Pair of RV dump valves										
•	1" PVC schedule 40 ball valve			x	x	each	$12.69				Potable water; hardware store
•	1½" PVC schedule 40 ball valve					each	$25.66				
●	2" PVC schedule 40 ball valve					each	$32.44				
•	1" PVC schedule 40 swing check valve		x	x	x	each	$13.78				
●	3" ABS schedule 40 backwater valve	x				each	$31.11				Sewer; plumbing supply
	Filters										
•	Pantyhose	x	x			pair	$2.50		4	*50*	Cover legs; drugstore
•	Airless paint sprayer bag	x	x			each	$7.00		1	*12*	Painting; hardware store
	Pumps										
●	Zabel p-se-12t (integral switch)			x	(x)	each	$289.00	17			Dewater basements, pump septic effluent, etc.; hardware store, Little Giant Pump Company,[s8] ReWater Systems,[s5] specialty stores
●	Little Giant 8E CIA RFS (integral switch)			x	(x)	each	$357.00	54			
●	Little Giant 5.5asp (integral switch)			x		each	$221.00	35			
●	Little Giant 5msp (needs float switch)			x		each	$161.00	15			
●	Float switch					each	$58.00				

3-way diverter valve

Choosing and Finding Parts

Greywater collection plumbing parts are mostly the same as those for standard collection plumbing. Table 4.3 shows most greywater parts (except Branched Drain distribution system parts, which are in Table 9.2). Within the family of drain-waste-vent plumbing parts, ABS plastic drainpipe works well. This system has a huge diversity of fittings and bends. The PVC sanitary drainpipe system is apparently equally diverse (many fittings are made from pouring PVC in the same molds used for ABS fittings). I've not personally worked with PVC, though. (If you are working in the non-industrialized world, see *PVC Sanitario*, p. 80.)

A few fittings from other plumbing systems are used in greywater systems; these are noted in Table 4.3. With those few exceptions, you don't want to use pipe and fittings designed for other purposes such as potable water, for example. (See Error: Freshwater Designs and Hardware Used for Greywater, Chapter 11.)

Minimize Plastic Impact

Design to use the *least amount* of the least damaging materials. HDPE is the least environmentally damaging plastic, followed by ABS, your most likely choice. PVC is the most evil, but hard to avoid. Cement works. Unfired clay or earth channels work well in the non-industrialized world. Other options aren't so practical: iron is rough inside and awkward to work with, copper is very expensive and damaging in its extraction. Bamboo would need frequent replacement.

Proper Fittings and Optimal-Size Pipe

For gravity flows, use sanitary drainpipe and drain fittings, which have no shoulder on the joints and have nice sweeping curves (Figure 4.6). Pressurized and/or filtered greywater can go through freshwater supply fittings.

All greywater plumbing should be corrosion resistant and of the proper diameter.

What size pipe and fittings should you use? Too big, and solids will form stationary islands in a thin film of water on the bottom of the pipe (not to mention the waste of plastic and money). Too small, and the pipe may fill to the top with water, overload, clog, and completely change the physics of flow splitting (Figure 4.7). A 2" pipe is plenty for the greywater main of any but the most palatial of residences and may take longer to clog. A 1½" pipe offers good scouring flows and economy.

FIGURE 4.6: SANITARY FITTINGS COMPARED TO SUPPLY FITTINGS

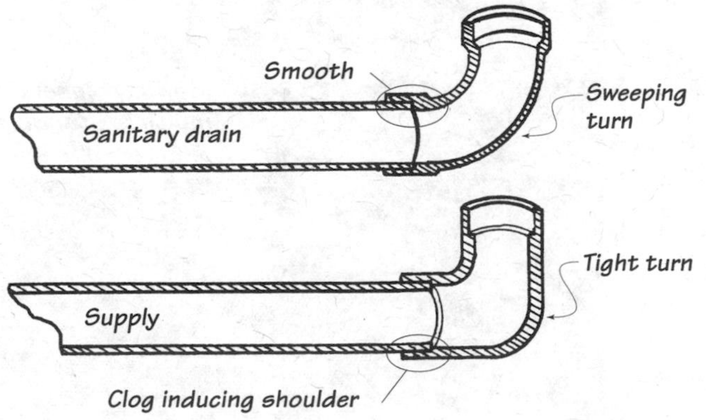

Smooth
Sanitary drain
Sweeping turn
Supply
Tight turn
Clog inducing shoulder

FIGURE 4.7: EFFECTS OF PIPE SIZE ON FLOW

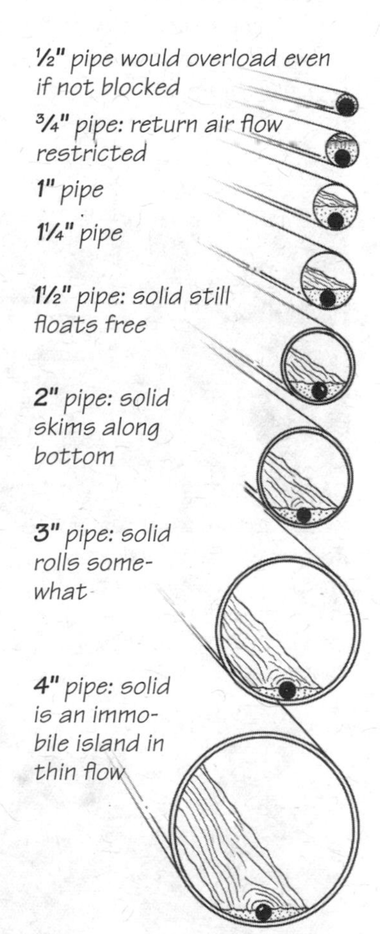

Valves

Valves are an expensive item on the parts bill. However, they are critical, so don't cut corners. (If you need an inexpensive system, consider one of the simplest, valve-less greywater systems, such as Landscape Direct or Radical Plumbing.)

Diverter Valves

You can't shut off the flow of greywater in the usual sense—if you generate it, it has to go somewhere. So the valves in simple greywater systems are generally *diverter valves*. They direct a flow to one destination or another. In collection plumbing, the diversion is from the landscape to the septic/sewer system; in distribution plumbing, it's from one irrigation zone to another, or to a disposal field. In most cases, you want to use a 3-way diverter valve. Other options include operating two or more shutoff valves together or moving a pipe or hose between different drain lines (see photos).

Ideally, unique portions of the flow have their own separate diverter valves. For example, I like to divert the kitchen sink separately so it can be sent to the septic/ sewer while the other greywater continues to go to the yard. You might want to do this at the first sign of system overload or if the yard doesn't need greywater at the moment, but you don't want all the greywater going into your septic tank. (Another option for a kitchen sink is to route the wash side permanently to the septic/ sewer and the rinse side to the yard.)

Note: 3-way valves are rare even in well-stocked plumbing suppliers—look for them in pool supply houses or order from oasisdesign.net.[s2,s9]

Simplest possible diverter valve: A coconut shell (rock, chunk of wood, dirt clod… whatever) is shifted by foot from blocking one greywater furrow to another.

Diverter for cheap labor conditions.
Slide the shroud shut, and water goes to trees. Slide it open, and water goes to treatment/disposal.
The shroud can also be set partly open, so low flows go to disposal and only high flows make it through to trees. This helps maintain aerobic conditions by providing more dry out time. This valve was made in rural Mexico of PVC sanitarioand cement. It could be done in ABS with an unglued wye, which rotates up to send the water straight or down to let it out.

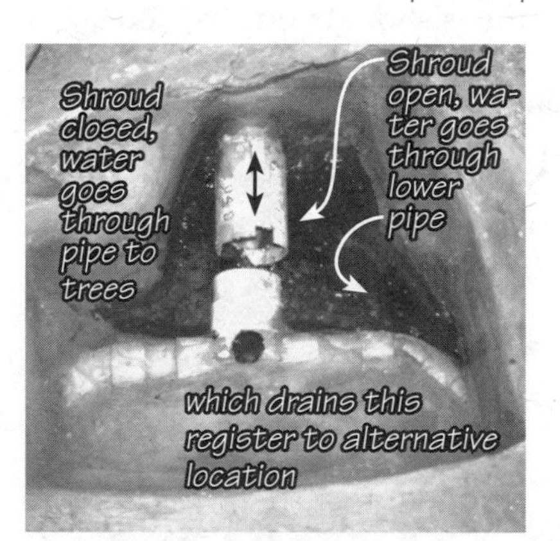

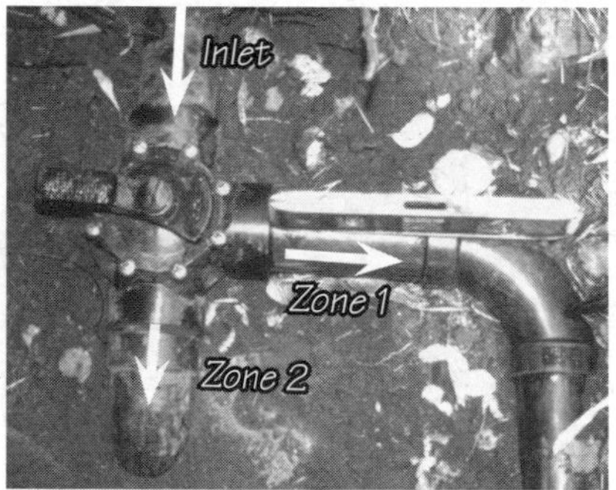

Brass 1" 3-way valve for washing machines.

Plastic 2" 3-way valve.
Interior of 3-way diverter valve would ideally be smooth, but it is not at all. It is one of the mysteries of greywater system design that these things clog so seldom, while so many other things clog so often.

Moving washer hose between multiple standpipes achieves diverter valve effect.

A 3-way diverter valve is spliced in under a sink. Note that the kitchen sink trap had to be raised to make space.

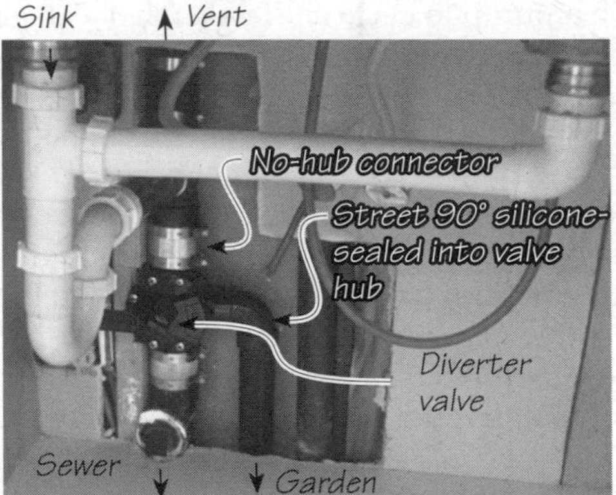

Diverter Valve Installation

The installation of 3-way valves to toggle between greywater destinations is likely to be unfamiliar but readily understandable to most plumbers.

The diverter valve must be accessible if you want it to get used very often. This is tricky with retrofits, as the valve must be positioned between the fixture vent and the septic/sewer connection, and all of this is usually in an awkward corner under the house. Sometimes you can reach the valve from a hatch in the floor, or through a cabinet under the sink. Some diverter valves can be fitted with an extension handle so that a valve installed under the house can be turned from inside the house. A motor drive is available for Jandy diverter valves. These can be controlled remotely with the push of a button (if you've got $450 for the actuator, and provided there is power).

Verify that access to turn the valve is as easy as possible, there is access to remove the face, and a bunch of fall is not squandered in its installation. If you can, route the greywater out of the house and put the valve there. If necessary, do a U-turn for the line connecting to the septic/sewer.

Note that the "inlet" on a Jandy 3-way valve can be changed to be any of the three ports of the tee by unscrewing and rotating the valve face. This handy feature opens more options for the design of this complicated junction. If the inlet is the leg of the tee, and you position the valve so both outlets are open, the flow will *split* into both outlets, though not nearly as cleanly as with a double ell. Diverter valves are best installed with no-hub connectors or silicone sealer (see Build for Future Flexibility, p. 23). Also, it's a nice touch to give pipes a tweak inside the hub so they come out of the valve sloped instead of flat (as in the photo "A 2% downward tweak on pipes," Chapter 10, p. 93).

VALTERRA[S23]

An RV dump valve is inexpensive, takes less space, and can be replaced without cutting pipe.

Two Shutoff Valves in Lieu of a Diverter Valve

In a pinch, two shutoff valves (below) work in lieu of a 3-way valve. However, if both shutoff valves are shut by mistake, an upstream pump can burn out, or water with nowhere to go may overflow. Also, the dead-end space before a closed valve goes anaerobic and accumulates crud (diverter valves don't pose these problems). If the space is deep enough and the valve shut long enough, the crud forms a solid plug so no water will pass when the valve is reopened. Two RV dump valves can be used; they take less space than ball valves, so the crud plug would be smaller. And, they disassemble to get the plug out.

Shutoff Valves

More complex greywater systems have shutoff valves that enable various functions such as draining a filter or surge tank, etc. Gate valves are less expensive than ball valves, but are far more likely to clog with hair or lint and fail to shut off completely. They are not recommended. RV dump valves work well. They are reasonably priced, and disassemble to provide complete access for service or replacement.

Check and Backwater Valves

Check valves prevent water in lines that slope uphill from running back into the surge tank and causing a pump to cycle endlessly. They can also be used to prevent blackwater from backing into a greywater line. Use *swing* check valves, not *spring* check valves, as the latter clog too easily. A fancy check valve called a backwater valve is designed for sewage.

Electric Valves

Most inexpensive electric solenoid irrigation valves don't work with greywater systems. There isn't enough pressure to operate their diaphragms, they corrode quickly, and soap sticks the diaphragms shut in greywater. However, a few do work. The Toro 2500t, for example, has a sensitive diaphragm that works at 15 psi and doesn't stick.

Very expensive motorized electric valves designed for wastewater are available, and Jandy sells motorized diverter valves for pools, which also work.

Tools for Collection Plumbing

The following tools are almost always necessary for collection plumbing:

- tape measure
- levels (for gauging correct slope)
- hacksaw, plastic pipe saw, or sawsall
- ABS glue
- drill and large bit or hole saw (to make holes for pipe to go through)
- ¼" nutdriver or wrench (for tightening no-hub connectors)
- adjustable wrench
- knife (to de-burr pipe ends)

And these tools are sometimes needed:

- transit or water level (for checking levels across long distances; see Appendix B)
- sawsall with metal cutting blade (to cut cast iron drainpipes used in most pre-1970 homes)
- cordless drill (for pipe hangers)
- narrow trench shovel (for running pipes underground to yard)
- pipe wrenches

Radical Plumbing: A Fraction of the Resource Use

This is the way to go if you are (or want to be) way ahead of the curve on living better with less resource consumption. You don't need or want a plumber if you go this route.

"Radical" means root. A radical solution addresses a problem—in this case, the drawbacks of septic and sewers—at its root.[15,16] The greatest benefits at the lowest cost are achieved when there is no redundant wastewater system—when a composting toilet/greywater system is not an add-on but replaces the septic/sewer entirely. That is the genesis of Radical Plumbing. Elimination of a redundant backup system has implications for greywater system design:

The distribution system needs to be designed to be able to handle all the greywater load, all the time. In practice, this won't change the design much. Most greywater-only systems are built in contexts with nonexistent regulations, where the perfection standard is simply shifted to accept occasional greywater system overload. By any rational analysis, greywater system overload is of little consequence, either absolutely or compared to septic or sewer system failure, so this isn't generally problematic.

Where a greywater-only system is permitted, the requirements for a permitted system generally include so much over-capacity that the absence of an alternative disposal system shouldn't increase the design requirements. On the other hand…

The collection plumbing can be *greatly* simplified. An interesting domino effect occurs in collection plumbing design when you commit fully to your greywater system and forget about a septic/sewer connection.

Drainpipes have water-seal traps to keep sewer gas out of the home. To keep the water from being siphoned out of the traps, the traps need to be vented. Wastewater rushing down a pipe into a septic tank displaces air. Because that air has no place to go other than out the vent, it has to rush back up the drainpipe even as the water rushes down. One of the reasons drainpipes are so darned big is to accommodate this air countercurrent above the water stream.

Well, if there's nothing but green garden and blue sky downstream from your greywater lines, you don't need gas traps, which means you don't need vents. Moreover, any displaced air can head out into the garden *ahead* of the water, so the pipes can be considerably smaller. Then (by gosh), once you're freed from the fancy-schmancy turns and extra connections for

FIGURE 4.8: RADICAL VS CONVENTIONAL PLUMBING

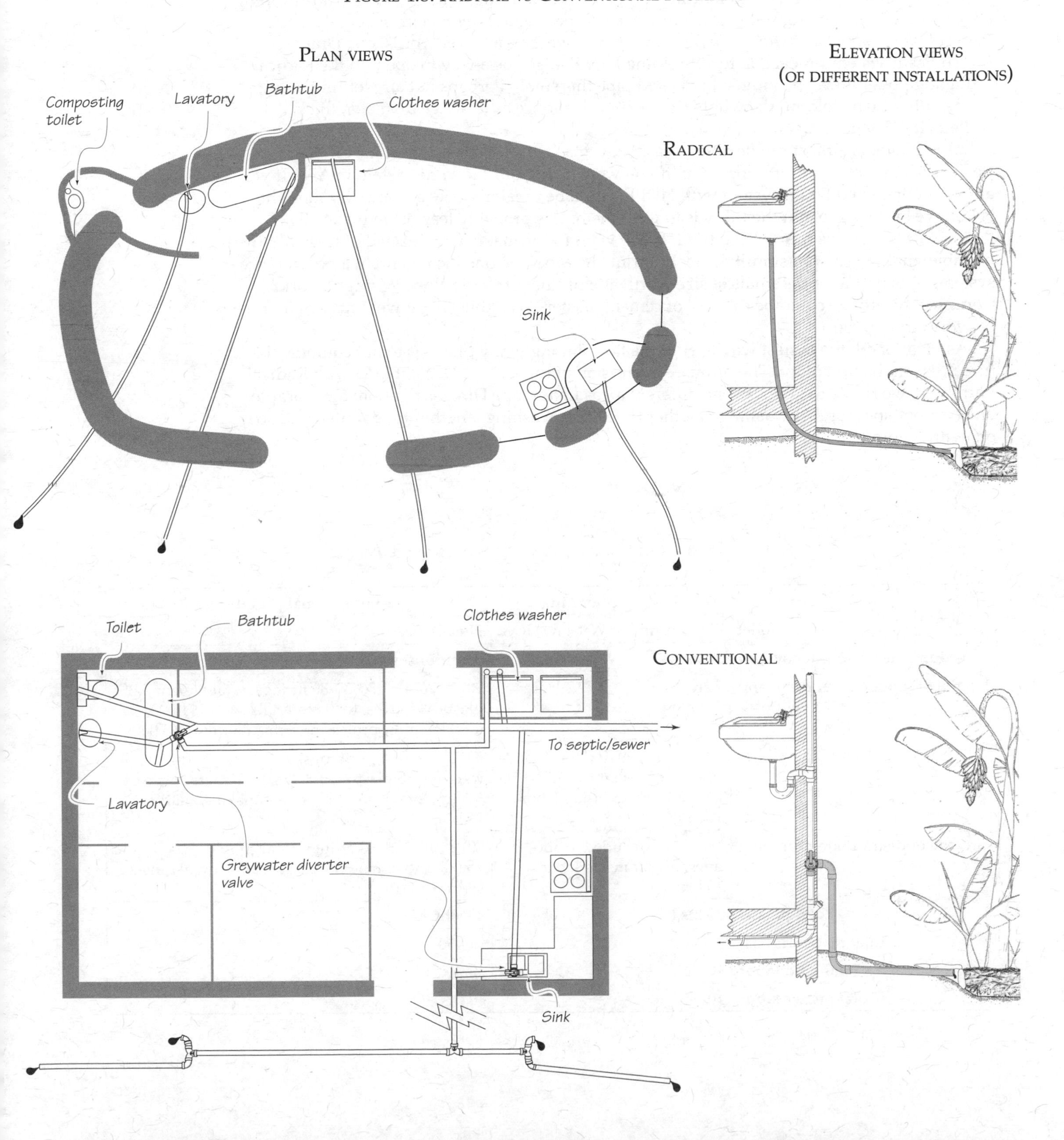

traps and vents, you can switch from ABS (environmentally questionable) to HDPE (environmentally benign but tough to make joint and tight turns with).

Bottom line: Instead of plumbing your house with a truckload of ABS that costs thousands of dollars to install, you can get your water out with about $50 worth of HDPE that you can install yourself. Instead of 1½" to 2" pipe, you can use 1¼" HDPE, or even 1". (If you never thought of this before, you've got plenty of company. Thanks to Ianto Evans for expanding my thinking on small, unvented pipes. *Note that you want to build a system that has no standing water if you have no traps, and that a drain screen is recommended to keep crawlers from crawling into your house through the pipe!*)

I am way into the HDPE pipe alternative. However, none of the fittings are suitable for drain water. Both the compression fittings and the barbed fittings leave ragged, crud-trapping edges. Therefore, you are pretty much restricted to making *one continuous run of tubing from each and every individual fixture to its own set of plants outside*—in other words, one Drain to Mulch Basin system for each fixture. Welding HDPE might be a way to open up the Radical Plumbing possibilities. It is unusual to weld pipe this small. Perhaps hot air welding might work, rather than welding the whole circumference at once as for large diameter, thick-walled HDPE pipe.

Also, HDPE is almost too flexible to easily lay at a perfectly consistent slope of 2%. Realistically, you want more slope than if you were laying out the same system in ABS. If you attempt a Branched Drain system with HDPE, you'll be breaking new ground. My thought is that the best way to join the tubes is to weld them. It is probably feasible to make HDPE flow splitters, but this hasn't been tried. If you want to try, make wyes instead of tees, and (if possible) make each split simultaneously shrink the pipe size one increment. In a no-pressure system, ¾" is probably the smallest size worth attempting. On a big slope ½" might work. Conceivably, fittings could be cast out of cement plaster. Available HDPE weld fittings might be worth trying, too.

As mentioned, the easiest way to split the flow among many plants is to not combine fixture flows in the first place. The ultimate expression of this principle, combined with Radical Plumbing, would be to give various different areas Landscape Direct irrigation: for example, one outdoor spot for dishwashing, another for clothes washing, another for bathing, and so on, with no pipes at all.

TABLE 4.4: RADICAL VS CONVENTIONAL PLUMBING

	Radical plumbing	**Conventional plumbing**
Toilet	Composting or Watson Wick[24]	Flush toilet
Redundant wastewater system	None	Septic or sewer
Length of septic/sewer drain and vent pipe	Zero	30' 2", 20' 1½", 30' 3", 80 fittings, total weight ABS plastic = 150 lbs, total materials cost = $1,000, total labor for professional installation = $3,000
Greywater collection plumbing	100' 1" line, 20' 1¼" line, no fittings, total weight HDPE = 30 lbs, $50, owner install labor = 10 hrs	30' 2", 20' 1½", 30 fittings, two 3-way valves, total weight ABS plastic = 100 lbs, total materials cost = $300, total labor for professional installation = $1,500
Greywater distribution plumbing	Materials included in above lines, owner install labor = 10 hrs	10' 2", 40' 1½", 30 fittings, total weight ABS plastic = 80 lbs, total materials cost = $150, owner install labor = 10 hrs
(2005 costs) **Total plastic**	30 lbs HDPE	330 lbs ABS
Total materials cost	$50	$1,450
Total paid labor	$0	$4,500
Total owner labor	10 hrs	10 hrs

Chapter 5: Greywater in the Landscape

Now that you've figured out how to get the greywater out of your house, it's time to consider how to handle it in the landscape. Key landscape design points were mentioned at the beginning of Chapter 3. Now we're going to get into more details.
A primary consideration is whether you are designing a system only to treat and dispose of greywater (in which case you can read the paragraph below and then skip ahead to Plants for Greywater Treatment/Disposal); or, whether you hope to reduce your freshwater consumption by using greywater for irrigation. In the latter case, we suggest you read this whole chapter carefully.

How Much Area Do You Need for Treatment/Disposal?

Treatment/disposal area is easy. Check the perk rate of your soil, and look up the needed area in Table 2.3, Disposal Loading Rates. Unless your greywater flow is huge, your perk terribly low, or your area really tiny, space shouldn't be a problem. It only remains to figure out how to distribute greywater over this area, using the greywater distribution system of your choice.

Coordinate with Freshwater Irrigation, Actualize Water Savings

Note outdoor faucets and existing freshwater irrigation hardware on your site map. Consider how greywater use will dovetail with freshwater irrigation to actualize your water savings. In theory, a main benefit of greywater reuse is saving freshwater. To achieve this in real life, you have to adjust your freshwater irrigation to account for the variable contribution of greywater. The easiest way is to have zones where freshwater irrigation can be independently reduced or eliminated. In an automated system, this is done with valved zones and adjusting the controller. In a manual system, it's done by paying attention.

In the ReWater automated system, makeup irrigation water is added automatically, and greywater in excess of plant needs (if any) is automatically shunted to septic/sewer.

Applying the permaculture credo "the problem is the solution," greywater and freshwater drip can be set up to form complementary functions. Eighty percent of the root mass of most plants is in the top foot or so of soil. A drip system can keep this area at optimal moisture for optimal growth. Then, occasional large doses of manually applied greywater will flush salts and encourage plants to send roots farther than little spots right under the emitters. Deeper roots provide drought insurance and make use of stored rainwater. If you have an automated greywater system with a predictable supply, the roles can switch, with the freshwater providing the occasional deep watering or a different wetting pattern.

Also, greywater can be directed to new plantings, so the entire drip system does not have to be turned on more frequently to meet the temporary extra needs of a few plants.

Irrigation Efficiency

The percentage of applied water actually used by plants is the irrigation efficiency. The rest evaporates from the soil or is lost below the roots. Drip irrigation delivers a precise amount of water to the targeted area(s) at the correct time. Eighty percent of the water from a well-designed drip irrigation system is transpired by plants. With few exceptions, the irrigation efficiency of greywater systems is low—50% or less.

The most intractable issue is that greywater generation is fairly constant, while irrigation needs may range from zero to many times the available greywater. Aside from that, the central challenge in greywater irrigation design, as with conventional irrigation, boils down to how to divide and distribute the flow.

The uniform, precise water distribution of drip irrigation is the result of relatively high pressure (15–20 psi, *100–140 kPa*) in a line with small orifices.[s10] Greywater, in contrast, discharges from the house at zero pressure, full of orifice-clogging glop.

It is possible to finely filter the glop out of greywater, pressurize the water, and distribute it via drip irrigation. Filters need frequent cleaning, a disgusting job soon surrendered by all except the truly fanatical. You can, however, purchase an automated, engineered solution to this problem—for a few thousand dollars or more.

In general, only for high flows or new construction can the cost and complexity be justified. (For more on pumping and filtration, see the pumped systems in the System Selection Chart, and Appendix E.)

Low-tech systems sacrifice irrigation efficiency for simplicity and economy. In the low-tech system example in Figure 5.1 (p. 40), 40% greywatering efficiency means only ⅓–½ of the total irrigation need is met by greywater, far short of what is possible in theory. But for most residential retrofit systems, this sacrifice is the best approach. In practice this means that you have far fewer outlets, with more water in each outlet. This in turn affects the choice of plants; with only a few outlets, it's a lot easier to irrigate trees than grass.

Choose the Proportion of Irrigation to Meet with Greywater

Greywater generation is fairly constant, and irrigation needs vary widely. The proportion of plants' irrigation needs you aim to fill with greywater affects how many plants to plumb to, how much effort to put into their supplemental freshwater irrigation, and the efficiency of the greywater reuse.

If you apply enough greywater to cover the peak need, most of it will be wasted most of the time. If you spread the greywater around so the application rate equals the lowest irrigation need, you will have to cover a wide area with greywater plumbing, and—unless you don't mind the plants being parched most of the time—freshwater irrigation plumbing as well.

In choosing among the options below, it can help to first draw out different scenarios

TABLE 5.1: POSSIBLE MATCHES BETWEEN GREYWATER DELIVERY AND IRRIGATION NEED

Greywater application equals ⇨	**Average wet season irrigation demand**	**Average dry season irrigation demand**	**Peak dry season irrigation demand or more**
Effectiveness of water reuse	Maximized	Medium	Very low
System overload	Very unlikely	Possible in wet season with adverse soil and rainfall conditions	Likely without carefully considered design
Greywater irrigation shortfall	Large; supplemental irrigation system required	Moderate; some supplemental irrigation required	Little, none, or year-round surplus; no supplemental irrigation system necessary
Infiltration area required per 100 gpd *(376 L)* **greywater (approximately a conservative 4-person household)**	500–2,000 ft² *(30–200 m²)* outdoors in dry climate	200 ft² *(20 m²)* outdoors	Inside solar greenhouse: 40–200 ft² *(4–20 m²)* Outdoor mulch basins: 200–800 ft² *(20–80 m²)* Outdoor wetland system: 100–400 ft² *(10–40 m²)*

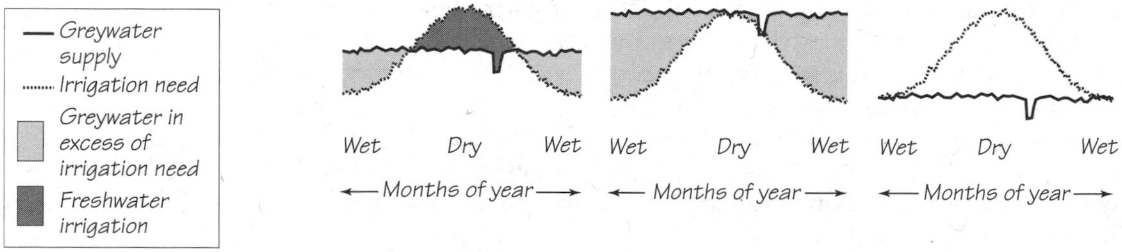

for different irrigation proportions on the site map—that is, scenarios in which the greywater is distributed among more or fewer plants. Here are the advantages and disadvantages of different irrigation proportions:

- **All else being equal, greywater delivery should equal the average dry season irrigation demand**—This ratio represents a good overall balance between irrigation efficiency and plumbing effort. This is what I aim to achieve with a typical residential system. Supplemental irrigation will be required in the driest times.

- **To avoid the necessity of an alternative irrigation source, greywater delivery should equal the peak dry season irrigation demand**—Much greywater is "wasted" in this scenario, as it is delivered in excess of plant needs. On the other hand, you save the effort of making any other irrigation system; it's all covered by greywater.
  Example: An institutional system with a large greywater flow irrigating a modest landscape through subsurface drip. Saving the cost of a freshwater irrigation system helps pay for the greywater system. The high efficiency of the drip irrigation effectively increases the greywater supply, helping ensure that the landscape is adequately watered. Also, with numerous, closely spaced outlets, each applying water at a very low rate, greywater is so evenly distributed that excess watering is less likely to create sodden conditions.
- **For disposal only, that is, for systems with little or no reuse, greywater delivery often equals or exceeds the peak dry season irrigation demand**—This ratio is most often found in wet and/or cold climates, or in areas with large wastewater flow in a small area. The primary goal is to get rid of the water in an ecologically sound way, with reuse after treatment or not at all. Plants can effectively treat an amount of greywater that is many times their irrigation need. However, applying excess water to ordinary plants in low perk soil would likely suffocate the roots. Wetland plants don't have this problem (see Constructed Wetlands, Chapter 8). It can also be reduced by making "high island" mulch basins. Putting the outlet a considerable distance from the trunk, possibly outside the canopy of the mature tree, would help too (see Mulch Basin Design, p. 47).
  Examples: Constructed Wetlands, sometimes soil beds in a greenhouse (purified excess water seeps out of drain tiles to reuse, or through the bottom of the beds to groundwater).
- **For maximum reuse, greywater delivery should equal the average wet season irrigation demand**—This is most attainable in dry climates, and with a fair amount of land (see Table 5.1 for area requirements). There isn't any point in designing a system to deliver *less* than the average wet season irrigation demand; just spread the water over fewer plants and save your plumbing effort.
  Example: Branched Drain to Mulch Basins, with evergreen fruit trees, in a desert (evergreens transpire at least some water in the wet season). Extensive supplemental irrigation will be required.

How Much Area Should You Irrigate?

Irrigation ability is limited when irrigation need is high but greywater generation is low.

I used to calculate irrigation need with the formula in our *Builder's Greywater Guide*.[6] Then I realized that this level of precision is highly desirable for people irrigating vast fields with freshwater, but useless for small-scale landscape irrigation with greywater. First, irrigation need can vary by a factor of 10 depending on whether the day is foggy, or hot with a dry wind. Moreover, what are you going to do about it? Greywater generation does not vary according to irrigation need. There's only so much laundry you can save for a hot day, only so many dishes you can wash (I suppose you could take more showers). The best you can do is match your greywater to some sort of seasonal average need for the landscape (as discussed in the preceding section), then make up any shortfall with freshwater.

Even if you use the coarsest of guesses for irrigation need, don't forget to account for tree growth. For a Branched Drain system, I suggest planning for how big the trees are going to be in ten years. That's a reasonable balance between current and future needs, considering the difficulty of changing the system and the likelihood that further in the future, everything else may have changed. For other systems, which have a shorter life or are easier to change, you could plan for fewer years of growth. How big will the trees be? Check out other trees in your area, ask a nursery, or research this at your library or on the Internet.

Take the irrigation need you calculated in Assess Your Irrigation Need, Chapter 2, and multiply it by the proportion you just chose in the previous section.

For example, if your irrigation need is 1,000 gpw for 2,000 ft^2 of tree canopy (per the rule of thumb in Assess Your Irrigation Need), and you have 500 gpw of greywater, you could:

- water all 2,000 ft² through two zones, alternating between freshwater and greywater irrigation in each zone during the high irrigation season, and watering all 2,000 ft² with greywater alone when irrigation needs aren't high
- water 1,000 ft² with greywater alone, with no backup irrigation system[m]

FIGURE 5.1: MATCHING GREYWATER SOURCES TO IRRIGATION NEEDS

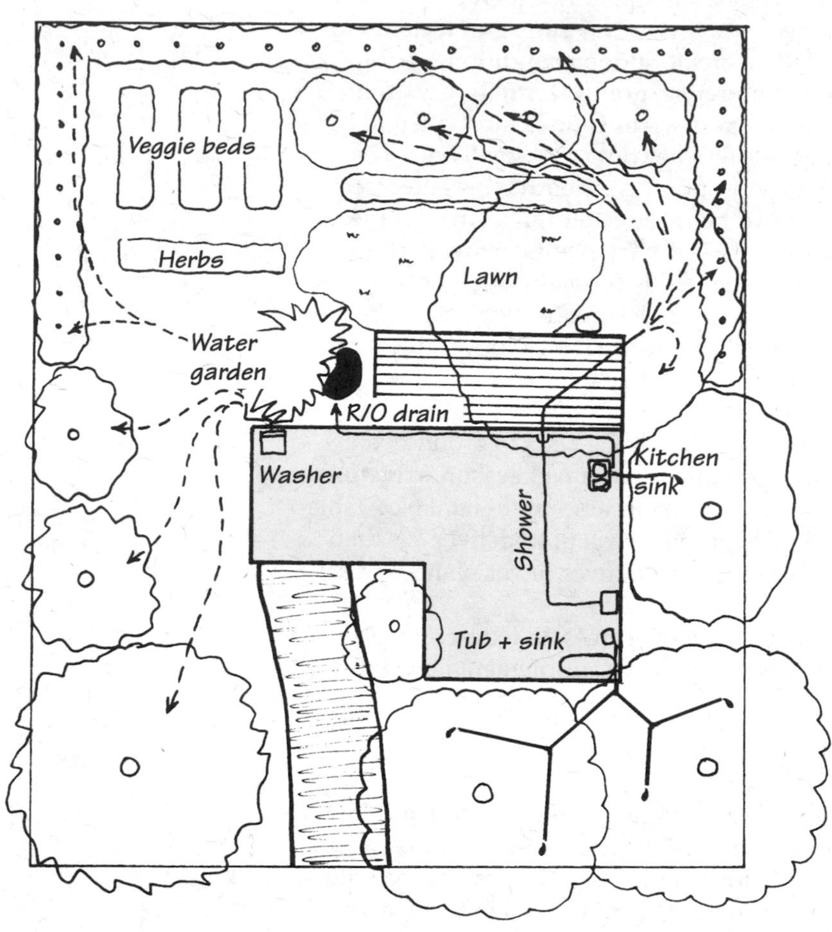

GREYWATER SYSTEMS USED:

- **Washer:** Laundry to Landscape system to downhill and slightly uphill plants.
- **Tub plus bathroom sink:** Branched Drain to Mulch Basins around large trees.
- **Shower:** Movable Drain to Mulch Basins, plumbed to opposite side of house, where irrigation need is greatest.
- **Kitchen sink:** Drain to Mulch Basin around one big tree.
- **Reverse-osmosis water purifier wastewater** (clearwater): Through ½" drip irrigation tubing to water garden. (Note: This small line would clog with greywater.)

% Irrigation Efficiency

Drip irrigation can reach 80% efficiency, about the highest attainable. The average irrigation efficiency in the example in Table 2.4 is 40%, so 864 gpw of reused water reduces freshwater irrigation by 350 gpw (25%).

Try different scenarios for connecting your greywater sources with irrigation need on your site map. Then do a mock-up using uncut, unglued pipe laid out on the ground. (See Figures 5.1, 9.4, and 12.1, and Table 2.4 for sample diagrams.)

Effect of Soil Type on Irrigation Design

The smaller the soil particles, the more water will spread in the soil, so the farther the emitters can be from each other (Figure 5.2).

Effect of Rainwater Harvesting and Runoff Management

Beside beneficially flushing salts, capturing rainwater runoff can reduce irrigation need. Account for this in your irrigation calculations.

FIGURE 5.2: EFFECT OF SOIL TYPE ON SOIL WETTING PATTERN

WETTED AREA APPEARING ON SOIL SURFACE

CROSS-SECTION OF THE WETTED AREA IN SOIL PROFILE

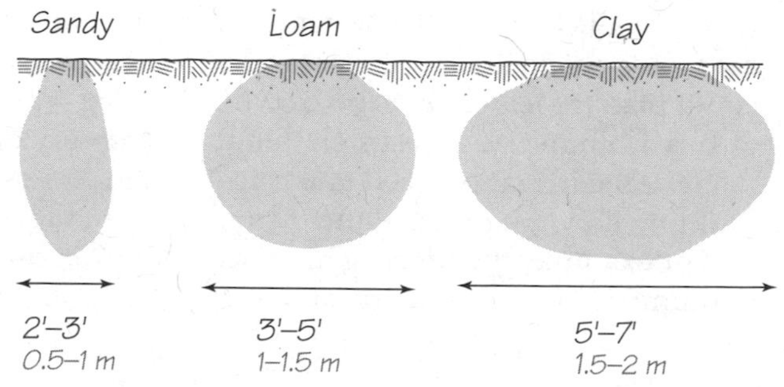

[m]*Metric:* *3,800 lpw irrigation need; 190 m² tree canopy; 1,900 lpw greywater.*

Cont. from p. 25

What to Do with Greywater When You Don't Need It

Is there any reason to route greywater to the yard when it's pouring rain? It depends. If you live where irrigation is not needed but wastewater disposal is difficult due to, say, low percolation or high population density, your greywater reuse may tend toward indoor reuse[34] or indoor cascading: using laundry rinse water for the next wash or using shower water to flush toilets, for example. (See The Household Water Cascade, Appendix F.) In dry climates, all the reasons to use greywater still apply until the soil has absorbed all of the moisture it can hold, whether it's raining or not.

Even when plants' water needs are met by rain, onsite greywater treatment is ecologically preferable to centralized treatment. Sewage treatment plants consume energy and chemicals. Their leftover nutrient-laden water pollutes surface waters instead of returning nutrients to the soil and recharging groundwater. During very heavy rainfall, runoff leaks into most sewage treatment systems and totally overwhelms their capacity. They release raw sewage into natural waters in these circumstances. People who don't know what to do with their greywater during heavy rain are in good company—the local treatment plant doesn't know either.

If your greywater would otherwise go to a treatment plant, getting it into the ground instead is beneficial almost no matter how wet the soil is. *(Exceptions: When soil is saturated, onsite greywater can flow over the surface, posing a potential health hazard. On an unstable slope, greywater could help oversaturate the soil, causing the land to slide. And if soil remains saturated for more than a day, plant roots can suffocate.)*

Greywater reuse can often continue under a blanket of snow (extreme cold is covered in Appendix C).

If you have the option of diverting greywater into a septic tank at times when soil is saturated, this is ideal; the treatment level is good, nearly all of the water returns to groundwater, and sewage plant loading is not increased.

Preserving Soil Quality

This is a consideration for all systems. The drier your climate, the higher the percentage of greywater irrigation, the higher the clay content of your soil, and the more fruit you eat from the greywatered area, the more critical preservation of soil quality is.

When you irrigate with greywater, you think more about what you put down the drain. Biodegradable means "breaks down," whereas biocompatible means the breakdown products are good for, or at least not harmful to, the environment.

A substance's biocompatibility depends on the environment to which it is being added. Phosphate is good for plants on land, but causes harmful algae blooms in the ocean. No-phosphate biodegradable detergents are designed not to hurt aquatic ecosystems, but contain salt; salt doesn't hurt the ocean or its plants, but is toxic to land plants and soil (Table 5.2).

Insensitive plants may be watered for years with greywater containing non-biocompatible cleaning products. However, once damage is finally evident, it can be a really big job to repair the soil—definitely a case where an ounce of prevention is worth a pound of cure.

Most of the substances in household greywater—for example, lint from natural fabrics, dead skin, sweat, hair, food particles, dirt, grime, etc.—biodegrade into plant nutrients. Household cleaners are the main exception. Also, salt from urine can be problematic with low rainfall and/or clay soil. Finally, for a society used to pouring any leftover liquid down the drain, the absence of a convenient means of toxic waste removal can be puzzling. We'll look at each of these issues in this section.

Garden-Friendly Cleaners

The chemical and biological composition of greywater varies greatly, based on numerous factors including the original quality of the water coming to the home, the habits of family members, the plumbing fixtures connected to the system, and the soaps used. After excluding more obvious toxins and unplugging the water softener, choosing plant- and soil-biocompatible cleaners is the major factor you can control to improve greywater quality.

Cleaner breakdown products vary in toxicity, and different soil types and plants vary in their susceptibility.

Continue p. 43

TABLE 5.2: BIOCOMPATIBILITY OF MATERIALS IN WASTEWATER WITH DIFFERENT DISPOSAL ENVIRONMENTS

<table>
<tr><th rowspan="3">Material</th><th colspan="4">Disposal Environment</th></tr>
<tr><th colspan="2">Terrestrial</th><th colspan="2">Aquatic</th></tr>
<tr><th>Arid land soil</th><th>Tropical and temperate soil</th><th>Ocean</th><th>Freshwater</th></tr>
<tr><td>Water</td><td>Highly beneficial</td><td>May be beneficial</td><td>No consequence</td><td>Little consequence</td></tr>
<tr><td>Organic compounds</td><td colspan="2">Beneficial food for soil microorganisms

Fast biodegradation times desirable</td><td colspan="2">Must be removed by oxidation in pre-treatment to prevent depletion of oxygen dissolved in water

Fast biodegradation times essential</td></tr>
<tr><td>N—Nitrogen</td><td>Extremely beneficial. N is beneficial plant growth's limiting nutrient.</td><td>Beneficial. May leach into ground-water if added in excess.</td><td>Probably OK. N is harmful algae growth's limiting nutrient, but dilution makes measurable effect unlikely.</td><td>Highly undesir-able. N is second most limiting nu-trient for harmful algae growth.</td></tr>
<tr><td>P—Phosphorus</td><td>Beneficial. Leaching into ground or sur-face waters unlikely due to low mobility in soil.</td><td>Extremely benefi-cial. P is beneficial plant growth's limiting nutrient.</td><td>Probably OK. P is harmful algae growth's second most limiting nutri-ent, but dilution makes measurable effect unlikely.</td><td>Extremely unde-sirable. P is most limiting nutrient for harmful algae growth.</td></tr>
<tr><td>K—Potassium</td><td colspan="2">Beneficial at washwater concentrations</td><td colspan="2">Effect unlikely</td></tr>
<tr><td>S—Sulfur</td><td colspan="2">Beneficial</td><td colspan="2">No consequence</td></tr>
<tr><td rowspan="2">Na—Sodium</td><td>Highly undesirable; toxic buildup likely</td><td>Undesirable but partly flushed from all but clay soils by rain</td><td rowspan="2">No consequence</td><td rowspan="2">Little consequence</td></tr>
<tr><td colspan="2">Directly toxic to plants, destructive to soil structure</td></tr>
<tr><td>pH—Acidity/alka-linity</td><td>pH lowering desir-able</td><td>pH raising desir-able</td><td>No consequence</td><td>Little consequence</td></tr>
<tr><td>Cl—Chlorine</td><td colspan="2">Undesirable</td><td>No consequence</td><td>Little consequence</td></tr>
<tr><td>B—Boron</td><td colspan="2">Highly toxic to plants at washwater concentrations</td><td>No consequence</td><td>Undesirable</td></tr>
<tr><td>Pathogenic micro-organisms</td><td colspan="2">Harmlessly biodegrade under proper conditions</td><td>Diluted but could spread disease</td><td>Likely to spread disease</td></tr>
<tr><td>Industrial toxins</td><td colspan="2">Disastrous</td><td>Highly undesirable; diluted but may bio-concentrate</td><td>Disastrous</td></tr>
</table>

Some cleaning products are toxic to plants, people, and the environment and should not be used. Products designed to open clogged drains or clean porcelain without scrubbing **must** be sent to the sewer or replaced with alternative products or boiling water and elbow grease.

Most hand and dish soaps and shampoos don't damage plants at low residential concentrations. Laundry detergents, on the other hand, need to be selected carefully. Powdered detergents and soaps include "filler" ingredients (not essential to clothes cleaning), which are usually some compound of sodium. Liquid soaps contain water as the filler, thus less sodium.

"Eco" claims on cleaners, even if true, apply to biocompatibility for aquatic disposal, and guarantee nothing about biocompatibility with plants and soil. Oasis and Bio-Pac soaps are formulated for use with greywater systems. (They are free of sodium, chlorine, and boron, and do not significantly affect soil pH.)

Few cleaners are so bad that they immediately kill plants. Water is a big plus for thirsty plants, toxins from ordinary biodegradable cleaners are a small (but cumulative) minus. Nutrients from biocompatible cleaners are a small plus. To minimize impact on plants and soil from cleaners, take these measures (in descending order of priority):

Effect of sodium-containing cleaners (center) vs biocompatible cleaners (left) *on tomato plants at high concentration. Control (no cleaners) is at right. (From a study I did at UC Berkeley, which led to the development of the first plant and soil biocompatible cleaners. I sold this business in 1996 and Oasis cleaners are now distributed by Bio-Pac.*[s11]*)*

- ❖ Avoid washing more often or using more cleaner than needed.
- ❖ Avoid cleaners that contain boron (borax), a potent plant toxin.
- ❖ Avoid using chlorine bleach or non-chlorine bleach containing sodium perborate, a potent plant toxin. Send bleaches to septic/sewer when used. Liquid hydrogen peroxide (e.g., Liquid Chlorox II or Ecover) is a less powerful, more expensive, but nontoxic alternative bleach. Do **not** use detergents "with bleach"; usually they contain sodium perborate.
- ❖ Avoid detergents that advertise whitening, softening, and enzymatic powers.
- ❖ Use cleaners that contain little or no sodium. Liquid cleaners and laundry detergents contain less sodium than powders. A buildup of sodium is toxic to plants and destroys the structure of clay soils.
- ❖ Use plant- and soil-biocompatible cleaners, which contain no sodium, chlorine, or boron and do not adversely affect soil pH or structure.[17,s11]

Portions of the information in the section above are reprinted with permission from the California Department of Water Resources Greywater Guide.[30]

Urine and Salt

The sources of salt in a greywater system are sweaty bathers, cleaning products, water softeners, and urine.

Water softeners often add high levels of sodium chloride that may adversely affect soils. Use of potassium chloride water softener salt[s1] makes softened water considerably less damaging. But—pulling the plug on softener is the best idea.

Urine is where the vast majority of the body's salt ends up. Urine may find its way into your greywater reuse system via toilets, a composting toilet leachate line, pee jars dumped into the system, or people peeing in the shower.

The good news on urine is that it is full of plant nutrients, primarily nitrogen, but also potassium and phosphate. The bad news is that it is equally full of sodium—it's half as salty as seawater. Each person whose urine goes into the soil adds 7 lbs *(3 kg)* of salt a year. Salt accumulates in soil, possibly to the level where it will kill plants. The rate of accumulation is greater the higher the soil's clay content and the lower the rainfall. If you have high rainfall and low clay content, urine would more likely be a longterm benefit than a liability to your plants, so you could route as much of it as possible into greywater. On the other hand, with clay soil and/or low rainfall, you're best off excluding urine as much as possible from your greywater system and routing it instead to the septic/sewer or to a disposal field.[18] A urine disposal field can be as simple as a deep post hole with a lid.

How do you know whether you've got enough rainfall to not worry about salt? If precipitation exceeds evapotranspiration—that is, if a barrel left in the open forever would overflow at some point in the year, rather than evaporate entirely—you've got a lot of rain. It's a continuum, but I relax a great deal about salt in climates that receive over 30" *(75 cm)* of rainfall annually.

The Key Role of Rainwater

In arid lands, freshwater irrigation raises the salinity and pH of the already too salty and alkaline soil. This simple, irresistible trend has been implicated in the collapse of all irrigation-based civilizations, including the impending collapse in the American West.[19] Unfortunately, greywater tends to do the same to an even greater degree.

This is a case where an ounce of prevention is worth hundreds of pounds of cure. To minimize the damage, don't soften water using sodium chloride, choose cleaners carefully, and direct urine appropriately for the soil and climate, as mentioned previously. In addition, **use concentrated rainfall and/or runoff to flush salts from irrigated areas.**

Rain (which has virtually no dissolved solids) is highly effective for flushing excess salts from the root zone. In rainy climates with free-draining soil, flushing is automatic. Otherwise, run rainwater from the roof downspouts or hardscapes into the mulch-filled basins that receive greywater. Give each basin several inches in a short period of time, then divert the rainwater to the next basin. Mulch helps too, by reducing direct evaporation from the soil. In the case of lawns, they can be graded perfectly flat, with a surrounding berm, and filled with concentrated rain runoff like a bathtub.

Monitoring and Repairing Soil

If your soil is damaged already, all is not lost. If you see salt burn on leaf tips, add gypsum and flush deeply with water according to instructions from your local nursery to remedy the soil damage. The application rates are high—hundreds of pounds of gypsum for a suburban lot.

If your plants seem to be suffering, test the soil pH. If pH is over 7.5, it will benefit from adding mulch, gypsum, and/or elemental sulfur to lower pH.

After repairing damage, kick your soil problem prevention program into high gear. (Gypsum can also be added at lower rates as a preventative measure.)

Toxic Waste Disposal

Obviously, you don't want to use a greywater system as a toxic waste disposal, even if you're accustomed to using the septic/sewer this way. So what do you do with the nasty stuff? Here are some options:

- **Don't let dangerous toxins into your home**—Reducing or eliminating toxins is far and away the preferred option. The Washington Toxics Coalition,[20] among others, has good suggestions for nontoxic alternatives to just about everything. In many cases the nontoxic alternative is cheaper and performs as well or better. Most of the "need" for toxic junk has been conjured into existence out of thin air, along with the successfully implanted assumption that the more toxic and expensive the product, the better. If you simply dismiss toxic products out of hand, you're not going to diminish your quality of life.
- **Route toxins into their own industrial re-cycles**—Used motor oil, for example, can be re-refined into perfectly good motor oil.
- **Send toxins to a toxic waste dump**—Check with local government/industry for options.
- **Route toxins elsewhere**—For example, by diverting them to the sewer, or burning them. The problem with this approach is that there isn't really such a place as "elsewhere." Even if *you* don't swim in or drink the water your waste contaminates, someone will.
- **Make a dedicated low-level toxin disposal field**—This can be very small, as the volume is typically miniscule: the occasional latex paintbrush washout, excess urine (see above), and whatnot. On the upside, you're taking responsibility for your own waste; on the downside, you're creating your own little low-level toxic waste dump.
- **Evaporate the liquid and landfill the solids**—Particularly if you live in a hot, dry climate, it may be feasible to dump nasty/undesirable liquids in a bucket, then landfill the minute amount of solids left after the liquid evaporates. This still has the "no elsewhere" problem, but less acutely.
- **Sell them**—I once cleaned all the household toxins out of my parents' house and had them cued up in front to take to the toxic waste disposal. We were also having a garage sale, and I'll be darned if people didn't swarm over and *buy* all those pesticides, old paint, drain cleaner...a powerful illustration of the "toxic = good" brainwashing mentioned previously.

Plants

If dispersal is your only interest, you can skip to the Plants for Greywater Treatment/Disposal section. Conversely, if you are only interested in reuse, you can read the Reuse section and skip the Disposal one. If your climate has both a disposal and a reuse season, read both.

Using greywater on ornamentals that don't require acid conditions is safest, followed by fruit trees. As mentioned elsewhere, greywater should not be used on vegetables (p. 105) or on the surface of lawns (p. 15, 104). Plants can be used to perform multiple system functions. For example, a solid ring or spiral of plants can provide privacy screen and shade for an outdoor shower or washing area, as well as absorbing excess water and nutrients.

If your soil is high perk and there is irrigation need in the wet season, you can probably greywater the same area year-round. Otherwise, you may need to make a separate zone for disposal during the wet season.

TABLE 5.3: SOME PLANTS FOR GREYWATER REUSE

1 = best
5 = worst

Warm climate ↔ Cold climate

Plant	Wet tolerance	Dry tolerance	Auto-mulch/barrier	Keeps leaves (transpires all year)	Salt tolerance	Benefits	Comments
Banana	2	3	3	Y	1	Delicious fruit, ornamental	Premier plant for greywater in warm climates. Clumps will expand until they are using all available water. If they need more water, chop a few down and the rest will do better. For maximum fruit production, each clump should have a mature, medium, and small stalk.
Mango	2	3	3	Y	3	Delicious fruit, ornamental	Grows into an enormous tree.
Avocado	4	4	3	Y	4	Delicious fruit, ornamental	Grows into an enormous tree. Make sure you have an idea of how to water it when it is four stories tall! Susceptible to root rot—use high island mulch basins (Figure 5.7).
Citrus	3	4	4	Y	2	Delicious fruit, ornamental	Great plant for greywater in subtropics. Highly fire-resistant, so good by the house.
Pineapple guava	3	2	4	Y	3	Delicious fruit and flowers, hedge, ornamental	Can be maintained as a 4' hedgerow, or shaped into a 20' specimen tree.
Fig	1	2	5	N	1	Delicious fruit, ornamental	No paradise is complete without figs. While deciduous, there is a shorter interval between dropping old leaves and growing new ones than for most deciduous fruit trees.
Clumping bamboo	2	3	4	Y	3	Structural wood, ornamental, slope stabilization	Wide variety of climates with different properties. Use clumping rather than running bamboos—much easier to contain or eradicate.[21]
Blackberry, elderberry, currants	2		2	N		Delicious fruit	Nothing like a stand of blackberries to attract people to the edge of your basin, but keep them out of the middle! Too bad most berries lose their leaves. Hard to eradicate.
Peach, plum, apple, pear, quince	4	4	2	N	4	Delicious fruit, ornamental	Good in cold climates; may require diversion of greywater at times to avoid waterlogging.

Plants for Greywater Reuse

Trees and shrubs are the best things to irrigate via low-tech greywater systems with few outlets. Water delivered anywhere in their root zones benefits the whole plant. It is a heck of a lot easier to hit a few big root zones than numerous small ones. Anything less than a large shrub is just too small for a system capable of supplying only a dozen or so outlets (as compared to hundreds with drip irrigation).

My philosophy is that if I water something, I want to be able to eat it. In my opinion, fruit trees are the logical thing to irrigate with greywater, especially in warm climates. For outdoor greywater systems in cold, severe climates, it's a different story. As you can see in Table 5.3, productive reuse options are mostly limited to plants that drop their leaves in winter. You may need a winter disposal zone, or forget about productive reuse and settle for ecological treatment.

If you've got fruit trees already planted, water them if you can. If you have evergreen fruit trees, they are the priority. They transpire and benefit from greywater year-round. If your trees are not yet planted, then you have the opportunity to optimize coordination between the greywater system and the layout of the edible landscaping.[22,s9]

TABLE 5.4: ACID-LOVING PLANTS
(DIFFICULT TO KEEP HAPPY WITH GREYWATER)

Azalea	Begonia	Bleeding Heart *(Dicentra)*
Camellia	Fern	Foxglove
Gardenia	Hydrangea	Impatiens
Wood Sorrel *(Oxalis)*	Philodendron	Primrose
Rhododendron	Zylosma	Violet
Blueberry		

Plants for Greywater Treatment/Disposal

For disposal, theoretically you don't even need plants. The bacteria on the soil particles can take care of pathogens by themselves (as in a sand filter). However, it's better with plants. They will remove the nitrate, phosphate, etc. from the water, and help maintain aerobic conditions.

In high perk soil with high water table, groundwater contamination could occur if wastewater moves through too fast. In very low perk soil, ponding and anaerobic conditions can occur. Plants improve physical soil conditions including perk rate, and add their own substantial contribution to water disposal by transpiration. A given area's treatment capacity is storage plus evapotranspiration plus percolation. Deepening mulch basins and choosing plants with higher water uptake can increase treatment capacity. The ideal plants for greywater disposal:

- are tolerant of wet conditions
- generate their own mulch or physical barrier so you don't have to (preventing greywater from being seen or played in)
- keep their leaves all year (so they transpire water all year)
- look beautiful and/or make fruit or some other useful thing (the plants in Table 5.3 work for treatment as well as reuse)

More Plants

There is a list of over 100 fruit trees for soft subtropical climates (reuse), and another list of over 100 plants for harsh, freeze-dried New Mexico conditions, at greywater.net.

TABLE 5.5: SOME PLANTS FOR GREYWATER TREATMENT/DISPOSAL

Note that all Table 5.3 plants work for treatment as well, particularly banana and bamboo.

1 = best
5 = worst

Plant	Wet tolerance	Auto-mulch/barrier	Keeps leaves	Benefits	Comment
Ivy, vinca, et al.	2	2	Y	Ornamental	Can be noxious scourges in the wrong spot, but do a good job of covering greywater basins so that added mulch is not needed.
Wetland plants	1	?	M	Pump oxygen into saturated soil	Most likely to be appropriate in cold and/or very wet climates.
Redwood	3	Y	Y	Timber	Conifers vary in wet tolerance, redwood can take a lot.
Willow, ash, cottonwood	2		N		All wet and cold tolerant.

If the perk is low and the loading rate high enough, you'll end up with a bog or wetland. Wetland plants literally pump oxygen down through their roots, maintaining aerobic conditions in mucky soil. For this reason, they are ideally suited for a heavily loaded treatment area. Also, many form such a dense stand that the need to add mulch is obviated; the plants themselves preclude access to the water (this matters because if there is even the tiniest area of open, standing water for more than a week or two, mosquitoes will hatch from it). For more on wetlands, see Constructed Wetlands, Chapter 8. For bogs, see the end of Chapter 10. To plant a small wetland, go to your nearest natural wetland and (with great care and conscientiousness) transplant *one or two* of each kind of plant that is abundant. Any of them appropriate to your greywater context will reproduce themselves in great abundance.

Infiltration Capacity of Greywater Mulch Basins

Septic leachfield infiltration capacity or longterm acceptance rate (LTAR) typically drops by 99% compared to adjoining soil, due to clogging by anaerobic bacteria.

Meanwhile, mulch basins typically have better *LTAR than the original soil.*

Mulch Basin Design

If I had to improve the world's handling of greywater in just two words, they would be **mulch basin.** Even wastewater flowing over the surface of the soil is purified to a surprisingly high level.[3.6] However, the simple measure of covering and containing the greywater in mulch basins assures a spectacularly high level of treatment. Contouring the ground helps contain runoff and concentrate irrigation water where needed, especially on slopes. Mulch basins and swales (long thin basins on contours across a slope, like terraces) are also perfect for capturing rainwater and storing it in the soil; swales on slopes and basins by downspouts.

Mulch

Mulch (ideally wood chips) *covers the greywater* so kids and dogs can't play in it. Greywater coming from a free flow outlet or hose quickly vanishes under mulch and is contained by the basin. Mulching softens soil's surface and slows the flow of greywater over it, allowing greywater to infiltrate much more quickly.

Mulch basins are a good way to reuse organic waste such as leaves, straw, prunings, weeds, and tree chips that might otherwise be landfilled. In many areas, truckloads of wood chips can be delivered for free (or close to it) by tree trimmers or road crews. Build your soil by tossing these reclaimed materials in mulch basins. Place the unchipped prunings at the bottom and the most attractive material on top. It will all biodegrade over time.

Mulch soaks up greywater, then time-releases it, which smoothes irrigation peaks and reduces erosion. Mulch slows the movement of runoff, retards evaporation (reducing salt buildup), helps lower pH, and provides habitat for beneficial organisms. Mulch reduces loss of water through gopher holes. Mulch adds beneficial organic matter and, eventually, nutrients to the soil. If you spread fertilizer on top, mulch binds it and then time-releases it. Mulch is basically *gold.* Spread mulch over all exposed soil in your yard and fill basins to the level of the surrounding soil.

Cautions: Dug-in mulch, which typically has a very high carbon/nitrogen ratio, can tie up nitrogen, starving plants. In areas to be seeded, avoid mulch from trees with germination inhibiting (allelopathic) properties, such as eucalyptus and black walnut. (This same property helps inhibit weed growth where desirable plants have already germinated.)

Basins

A **basin** is formed by scooping a doughnut-shaped hole in the earth and piling the tailings from the hole in a ring around the basin. Tailings can be piled downhill to make a basin on a slope. The walls of basins tend to get worn down, so they should be dug deep initially—10" *(25 cm)* at least—then filled with mulch. Size them to accommodate the surge volume they will receive: a bathtub and washer cycle's worth, for example (Table 4.2 gives surge volumes). The basin should be at least as wide as the dripline of the tree (the diameter of the canopy) and can be much wider. Most tree roots extend farther than their branches. Making the basins bigger than the tree makes it easier to expand them without hurting the roots as the tree grows (Figures 5.3, 5.4).

The basin *contains the greywater* where it is needed, within the dripline, and prevents it from escaping where it would be wasted or a nuisance. Equally important in wet climates, the rim around the basin *excludes runoff* from entering the basin, helping to keep the roots

FIGURE 5.3: STANDARD MULCH BASIN FOR NEW PLANTING

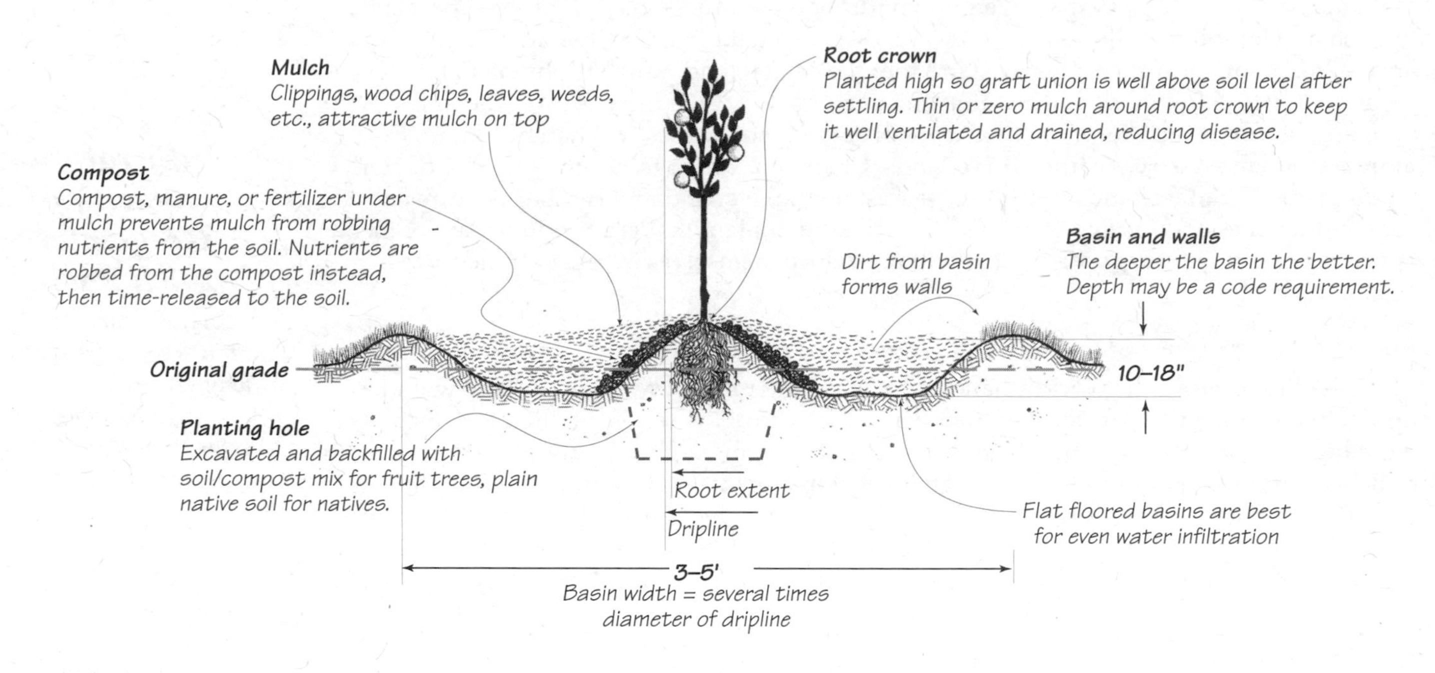

FIGURE 5.4: MULCH BASIN MAINTENANCE AND SAME TREE 30 YEARS LATER

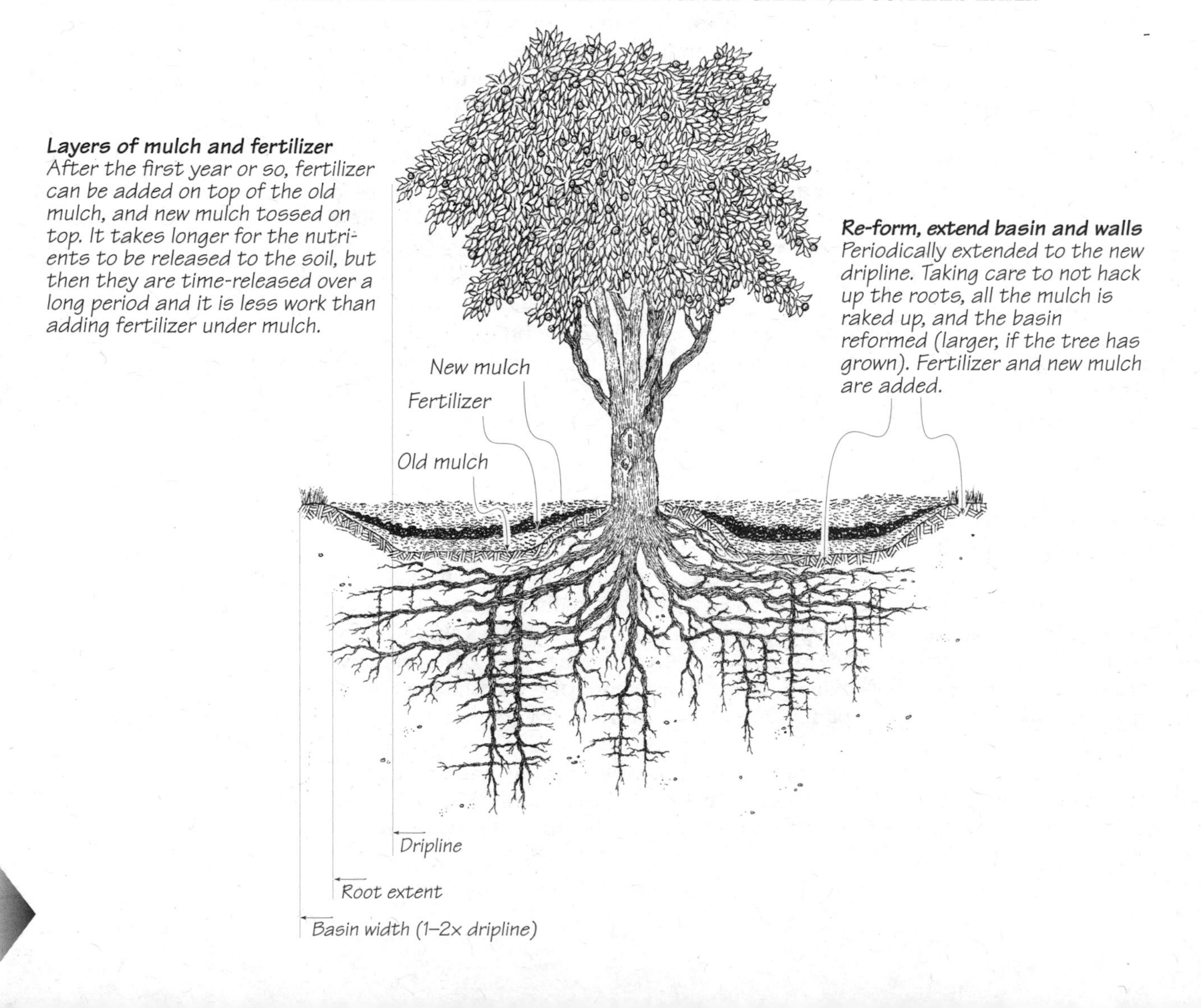

Layers of mulch and fertilizer
After the first year or so, fertilizer can be added on top of the old mulch, and new mulch tossed on top. It takes longer for the nutrients to be released to the soil, but then they are time-released over a long period and it is less work than adding fertilizer under mulch.
Re-form, extend basin and walls
Periodically extended to the new dripline. Taking care to not hack up the roots, all the mulch is raked up, and the basin reformed (larger, if the tree has grown). Fertilizer and new mulch are added.
New mulch
Fertilizer
Old mulch
Dripline
Root extent
Basin width (1–2x dripline)

Figure 5.5: Mulch Basin Terrace or Swale on a Slope

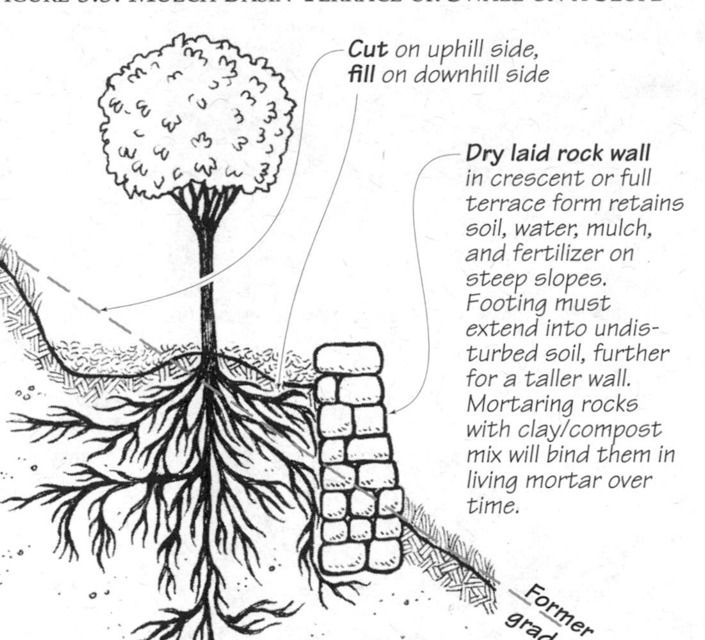
Cut on uphill side,
fill on downhill side
Dry laid rock wall
in crescent or full terrace form retains soil, water, mulch, and fertilizer on steep slopes. Footing must extend into undisturbed soil, further for a taller wall. Mortaring rocks with clay/compost mix will bind them in living mortar over time.
Former grade

Figure 5.6: "No-Island" Mulch Basin
For Water-Loving, Rot-Proof Plants (e.g., Banana)

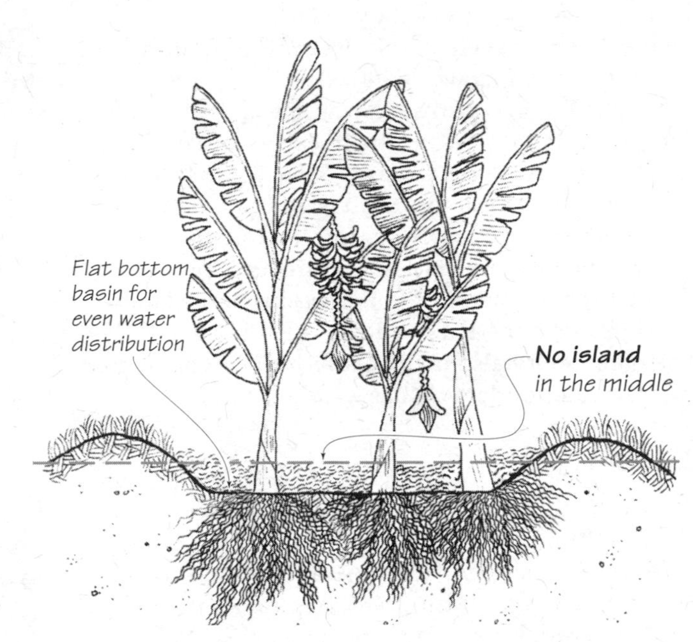
Flat bottom basin for even water distribution
No island
in the middle

Figure 5.7: "High-Island" Mulch Basin
For Water-and-Rot-Sensitive Plants (e.g., Avocado), and/or Plantings in Heavy Clay Soil

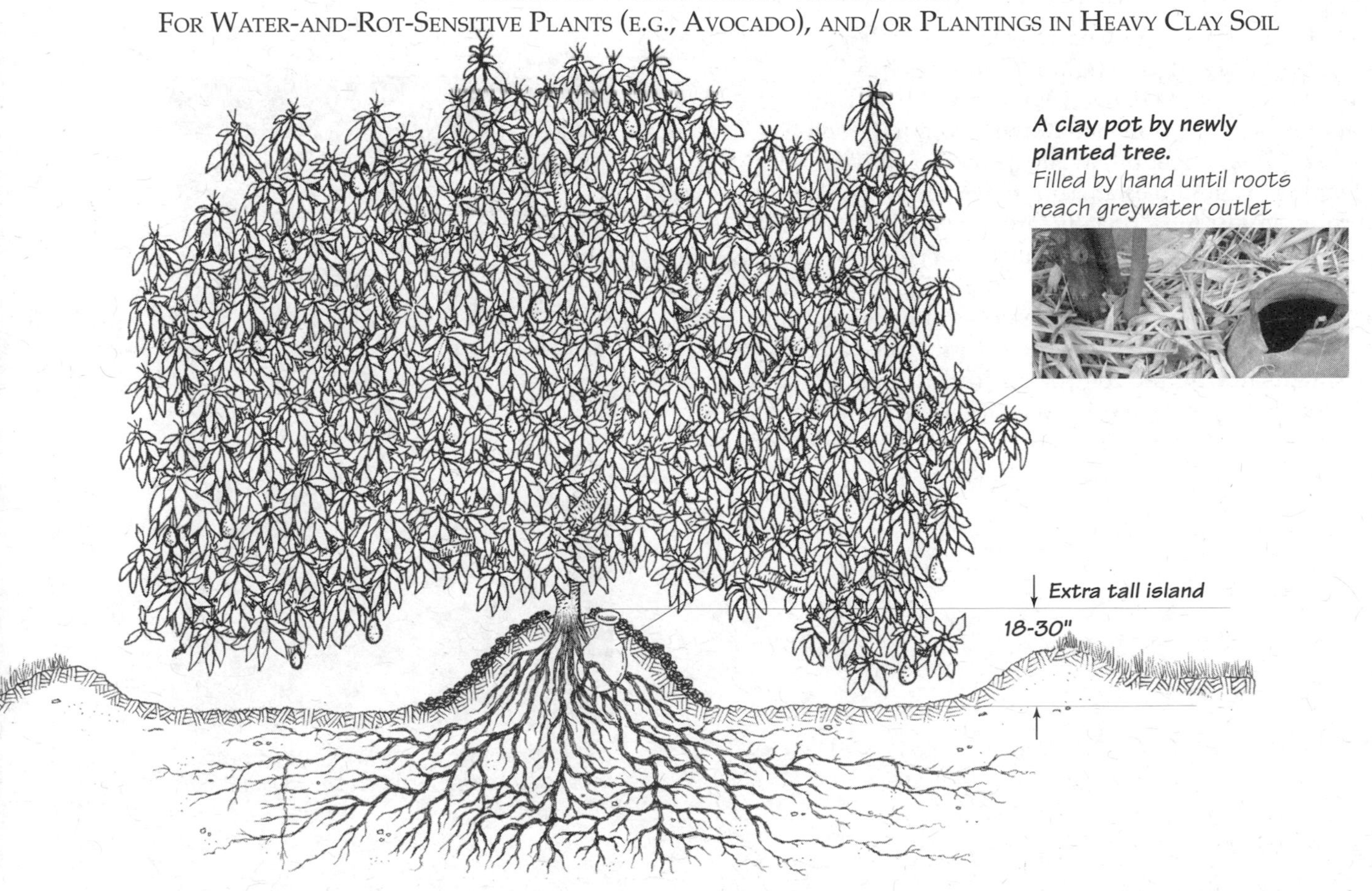
A clay pot by newly planted tree.
Filled by hand until roots reach greywater outlet
Extra tall island
18-30"

from drowning or the greywater from escaping in a flood of runoff. The island in the middle of the mulch basin protects the sensitive root crown from becoming waterlogged and diseased. In low perk/ high rainfall conditions the mulch basin may need an overflow, or even a drain on the downhill side (see Real World Example #4).

Mulch basin construction is especially critical for greywatered trees, as the trees will often get too much or too little water. Most trees like to be on the edge of the basin or on an island in the middle. Plants resistant to waterlogging can inhabit the bottom. See the mulch basin figures and photos for specific geometry of mulch basins and islands for different trees, soil, and climate conditions. The basin geometry also may be affected by what kind of greywater outlet you use.

Newly formed basin with tree.

Deep area of mulch basin adds surge capacity.
Peach tree
Greywater outlet; 1/8 of flow
"Shrubber" adjustable drip irrigation emitter can be shut off when greywater is available.

Basin with mulch, 2 years later *(note tree growth).*

Swales

Swales are long, thin basins (or blind-ended ditches) that run on contour, that is, on the same level across a slope. Their downhill tailings are an ideal place to plant fruit trees. (Figure 5.5 could be a section through a swale with a tree on the tailings.) Roots can seek water in the bottom of the swale during drought and air in the top of the tailings during flood. As with basins, swales should be filled with mulch.

Post Holes and Auger Holes

Post holes are like tall, narrow "basins." They are particularly suited to tight locations such as under a downspout between a house and walkway. Auger holes can be punched down through the soil with a 2" *(5 cm)* auger bit, available in 2'–5' *(0.5–1.5 m)* lengths. Auger holes amount to very little in terms of volume, but provide a proportionately large surface area for water to soak into lower layers of soil. These holes can completely change the infiltration characteristics of a soil underlain by a thin hardpan. They also allow roots to reach reserves of water and nutrients they could not otherwise. Fill them with mulch or compost. With plants nearby, the composting organic matter will turn into roots that help keep the soil open and infiltrating longterm. In Australian hardpan, dynamite has been used to create fracture patterns that fruit tree roots then follow.

Greywater outlet

Mulch basin with fertilizer added over mulch. *Nutrients bind to the mulch, then time-release.*

Chapter 6: System Selection Chart

Greywater systems consist of collection plumbing in the house, distribution plumbing, and the receiving landscape (outlet chambers, soil, and plants). In the next three chapters we'll help you decide which systems are right for your context.

Keeping in mind your context and the design considerations mentioned earlier, peruse the System Selection Chart that follows, with an eye for a system that meets your needs and your site. Take some time with this chart; it is your best tool for choosing a system.

A complete system provides for greywater collecting, distributing, and receiving. Some of the systems in the chart are complete, others are subsystems that can be combined with each other or with generic collection plumbing to make a complete system.

It is likely that at least one of the systems here will be adaptable to your situation. If not, each system's elements can be rearranged to make a wide variety of other possible systems, or to serve as an inspirational starting point for an as-yet-unconceived variation.

Note that some designs are cautioned as being unproven. Clearly, with experimental designs, the risk of wasting time and money is higher, so proceed with caution. An incremental approach can reduce this risk. For example, one fixture set could be connected to a single hose for a Movable Drain to Mulch Basins system. If everything worked well, the remaining fixtures could be connected, and then a Branched Drain system added later.

For best results, choose the simplest possible design and build it as well as you possibly can.

The next chapter describes the most simple greywater system options. Simple systems have the advantage that they work. The overwhelming majority of systems that are more than 5–10 years old and still working are very simple. Simple systems are ecologically benign, as they take little material to construct, no energy to operate, and generally last longer.

On the other hand, the more complicated and expensive systems in Chapter 8 offer more efficiency or convenience. All have applications, so they're all worth checking out.

Consider each set of fixtures individually; it may be best to employ a few different greywater systems, even in a small house. Quite often the laundry gets its own system, as does any fixture set that, because of collection plumbing issues, is not practical to unite with the other greywater flows.

Several years ago I developed the Branched Drain greywater distribution system in an effort to achieve some of the performance of a complicated system with the low cost and high reliability of a simple system. Branched Drain systems can automatically deliver greywater to plants without using a surge tank, filtration, pumping, or any moving parts. People can make this system themselves with off-the-shelf components.

Within a few years I found that I recommended the Branched Drain system for about 75% of my clients. I ended up writing a whole book about it in 2000. After selling several thousand copies, I began to feel that everyone who bought *Create an Oasis* should have the *Branched Drain Greywater Systems* book too—and now you do, because that book has been completely incorporated into this edition of *Create an Oasis*.

Chapters 9 and 10 cover Branched Drains in great detail. Even if you choose a different kind of system, you may find it useful to look at all design issues through the lens of the Branched Drain system.

The Laundry to Landscape system, a promising new washer pressurized greywater system, has been added to this 2009 edition, in place of the Drumless Laundry system, to which it is superior in every respect. If they continue working as well as they have so far, Laundry to Landscape systems will outstrip the Branched Drains in wide applicability. In particular, these hold promise as the first simple greywater systems that can be professionally retrofitted economically. We're promoting this business model as a free open source franchise. See oasisdesign.net/greywater/laundry for design updates.

Table 6.1: System Selection Chart

Legend:
m = maybe   1 = most satisfactory
y = yes   5 = least satisfactory
Y = yes, very much so
blank = no (or not applicable)

System Name	System provides for: C = greywater collecting, D = greywater distributing, R = greywater receiving	Overall score in optimum application	Renter's construction	Can distribute upslope of greywater source? D only	Collection possible with fixture plumbing in slab? C only	Suitable for non-industrialized world?	Proven?	Year current version developed (as far as we know)	Possibly legal under CPC/UPC? (All are legal in AZ, NM, and TX)[4]	Ease of construction (1 = easiest, 5 = hard)	Ease of use (1 = easiest, 5 = hard)
Simple, Easy Greywater Systems											
Landscape Direct	CDR	**1**	m		y	**Y**	y	BC	m	1–5	1
Drain to Mulch Basin/Drain Out Back	CDR	**1**	m			y	y	≤1990s		2	1
Movable Drain	DR	**2**	m			m	y evolving	1996*		2	3
Branched Drain	DR	**1**				y	y evolving	1998*	y	4	1
Laundry Drum	CDR	**2**	y		y	y	y	≤1970s		2	3
Laundry to Landscape	CDR	**1**	y	3 ft/ *1 m*	y	m	new, evolving	2008*	y	2	1
Garden Hose through the Bathroom	CDR	**3**	y	m	y	y	y	≤1970s		1	5
Dishpan Dump/Bucketing	CDR	**1**	y	y	y	y	y	BC		1	5
Mulch Basins	R	**1**	y			y	y	BC	y	1	1
Greywater Furrow Irrigation	CDR	**1**	y			**Y**	y	≤1995		2	3
More Complex Systems											
Drum with Effluent Pump	D	**4**	m	y			y, but mixed	≤1980s	y	3	4
Mini-Leachfields	R	**4**					y, but mixed	≤1980s	y	3	3
Subsoil Infiltration Galleys	R	**1**					y	1987	y	4	2
Solar Greywater Greenhouse	R	**1**		m			y evolving	≤1990s	y	5	2
Green Septic: Tank, Flow Splitters, and Subsoil Infiltration Galleys	CDR	**2**			y	m	y evolving	2000*	y	4	1
Constructed Wetlands	D(R)	**2**			y	m	y evolving	1900±	y	5	2
Automated Sand Filtration to Subsurface Emitters	DR	**1**		y			y, for high end only	1992	y	5	2
Septic Tank to Subsurface Drip	CDR	**2**		y	y		y evolving	1988	y	5	2

*Design presented here is original to Oasis Design, as far as we know

Both pages

Comment	Greywater Sources										
	Laundry, dishwasher	Tub, shower, bathroom sink	Kitchen sink	Toilet (blackwater) capable	Cost range *(for the functions that the system performs; add $200–1,000+ for collection plumbing if system is D or R only) (2005 costs)*	Requires electricity?	Adapted for freezing?	Suitable for large flows?	% Distribution efficiency (Drip = 80%) D only	Degree of filtration provided or required	Pages, refs to read (b = Builder's GW Guide)
Good for earthy lifestyle		y			$0–$1,000+				20–40[a]	None	54, b
Very simple, easy, reliable, cheap	y	y	y		$5–$10		y		10–40[a]	None	55, 74
Simple, good temporary system	y	y	y		$50–$100		y		10–50	None	56
Simple, reliable; most recommended system	y	y	y		$50–$1,000+		m		30–60[a]	None	74–100, b
Simple, very popular for laundry only	y	y			$20–$100	b			10–50	None to coarse	57
Simple way to water slightly uphill or on level, most recommended	y	y			$50–$500[c]	b	m		10–50	None	58–60[41]
Good place to start		y			$0				10–50	None to coarse	61
Good place to start	m	m	y		$0				30–80	None	61, 105–106, 114–115
Receiving landscape for most simple systems	y	y	y		$0		m		20–60	None	47–50
Very simple, easy, reliable, cheap	y	y	y		$0				20–60	None	127–133
Simplest system for uphill irrigation	y	y	m		$300–$1,300	y			20–50	None	24, 62, 133–134
Only in here because it's in CPC/UPC—Mulch Basins better	y	y	m		$200–$500		m	m	30–50[a]	Coarse to medium	63
Safe, requires batch dosing or Branched Drain	y	y	y	y	$100–$100		y	y	10–30[a]	None to medium	64
Best system for cold climates; an asset to a cold climate home in many ways	y	y	y	m	$100–$300 +greenhouse		Y	y	10–70[a]	None to fine	65–67[25,26]
Very promising, unproven system	y	y	y	y	Septic tank +$200–$3,000		y	y	20–30[a]	Fine (by settling)	67[24]
Most common in wet climates, familiar to regulators, expensive	y	y	y	y	$300–$15,000			Y	10–50	Coarse to fine	68[27]
Elaborate mechanics soothe health officials; proven safe for watering turf	y	y	y		$2,000–$6,000	Y		y	60–80	Very fine	69–70
Elaborate mechanics soothe health officials; proven safe for watering turf	y	y	y	y	$5,000–$30,000	Y	y	Y	10–50	Super fine	71–73

[a]Water path cannot easily be changed
[b]Electricity needed for washer pump
[c]For professionally installed, turnkey system

Chapter 7: Simple, Easy Greywater Systems

These proven systems are the simplest, and easiest. Each has its pros and cons—see if one fits your needs.

Each system has a code for the greywater function(s) it performs: C = Collecting, D = Distributing, R = Receiving. Most of the simple systems are complete (CDR); some are subsystems (DR, for instance) that need to be combined with whatever piece is missing (C, for instance) to make a complete greywater system.

Simplest of all greywater systems: Bath under hose in garden.

Landscape Direct (CDR)

Often the best choice for greywater systems in mild climates and in the non-industrialized world. I personally love *these systems.*

Perhaps you don't need to plumb the water through the house in the first place. Landscape Direct greywatering can be as simple as using the garden hose to take a shower under a fruit tree; it can be as beautiful as an elaborately landscaped bathing grotto amidst an edible jungle. This greywater may slip through a legal loophole in some areas: Because the water never enters a drain, it may not fit the definition of wastewater.

The most refined Landscape Direct systems use topography variations and materials with special permeability characteristics (such as sand or clay) to direct runoff efficiently to carefully sited plants.[22] Beside their other appealing features, these systems reduce the cost of a structure in which to bathe. Properly done, the treatment level is spectacularly high because unlike any other greywater system, there are no pipes or dark anaerobic corners in which pathogens could survive and multiply. These installations are especially nice inside a solar greenhouse in colder climates.

A Landscape Direct system would most likely be designed and/or built by a landscaper, as part of an Eden-like bathing grotto, for example (Figure 7.1). Construction of a bathing garden is part artistry, part intimate knowledge of local conditions and materials. To work out optimally, some science should go into it as well. The main considerations are when and how much greywater will be generated and how to bring it to the plants, or the plants to it. Ideally it should all be made out of natural materials, with as much as possible coming from onsite. As it is crucial that the system be adapted to local conditions, the variation in design from place to place will be extreme. One bathing garden may be a place to cool off, another to warm up. One design may take great pains to prevent waterlogging, another to concentrate water as much as possible.

Figure 7.1: Bathing Garden

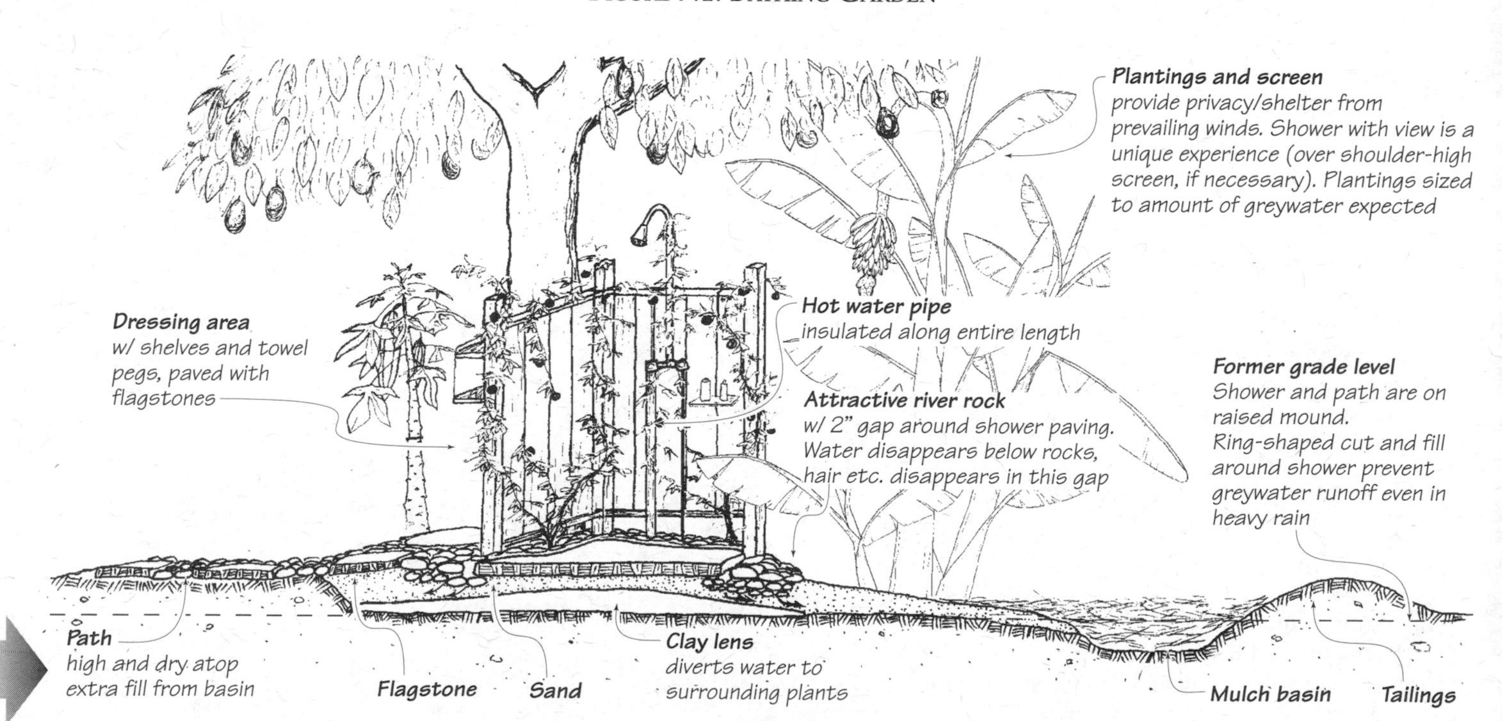
Plantings and screen
provide privacy/shelter from prevailing winds. Shower with view is a unique experience (over shoulder-high screen, if necessary). Plantings sized to amount of greywater expected
Dressing area
w/ shelves and towel pegs, paved with flagstones
Hot water pipe
insulated along entire length
Attractive river rock
w/ 2" gap around shower paving. Water disappears below rocks, hair etc. disappears in this gap
Former grade level
Shower and path are on raised mound.
Ring-shaped cut and fill around shower prevent greywater runoff even in heavy rain
Path
high and dry atop extra fill from basin
Flagstone
Sand
Clay lens
diverts water to surrounding plants
Mulch basin
Tailings

In general, the paths should be raised, and paved with something that you can walk over without getting your feet dirty. Likewise, the immediate shower area should be well drained. Water can be directed above or below ground by using materials of varying permeability. To get maximum irrigation utility from a shower on free-draining native soil, for example, a low-permeability clay lens can be poured under the surface of the shower so the water doesn't just fall straight away to the center of the earth, but is directed to a ring of plantings. Privacy and shelter from wind should be provided. (If you make a bathing garden, please send us a photo.)

Shower in Fiji
directly waters bananas, cocos, taro, etc.

Drain to Mulch Basin/Drain Out Back (CDR)

The Drain to Mulch Basin is an improved version of the Drain Out Back, the simplest system for just getting rid of greywater.

Probably 90% of the greywater systems in the world are no more than drains that point out the back of the house. Some are gross, and most don't reuse the water for irrigation. The simple refinement of adding a mulch-filled basin or sloping channel where the pipe dumps eliminates most grossness. Cultivating plants there whose irrigation needs match the water source can efficiently reuse the water. Cover the greywater outlet with rocks and mulch, and install a screen over the drain (or a vent and trap) unless you want the true backwoods feel, complete with vermin entering the house via the drainpipe. The lines are installed the same way as Branched Drain distribution lines (see Laying Pipe, Chapter 10). Lines can run any distance with continuous downhill slope.

Caution: In clay soils, applying greywater too near the house foundation may cause problems.[12]

FIGURE 7.2: DRAIN TO MULCH BASIN

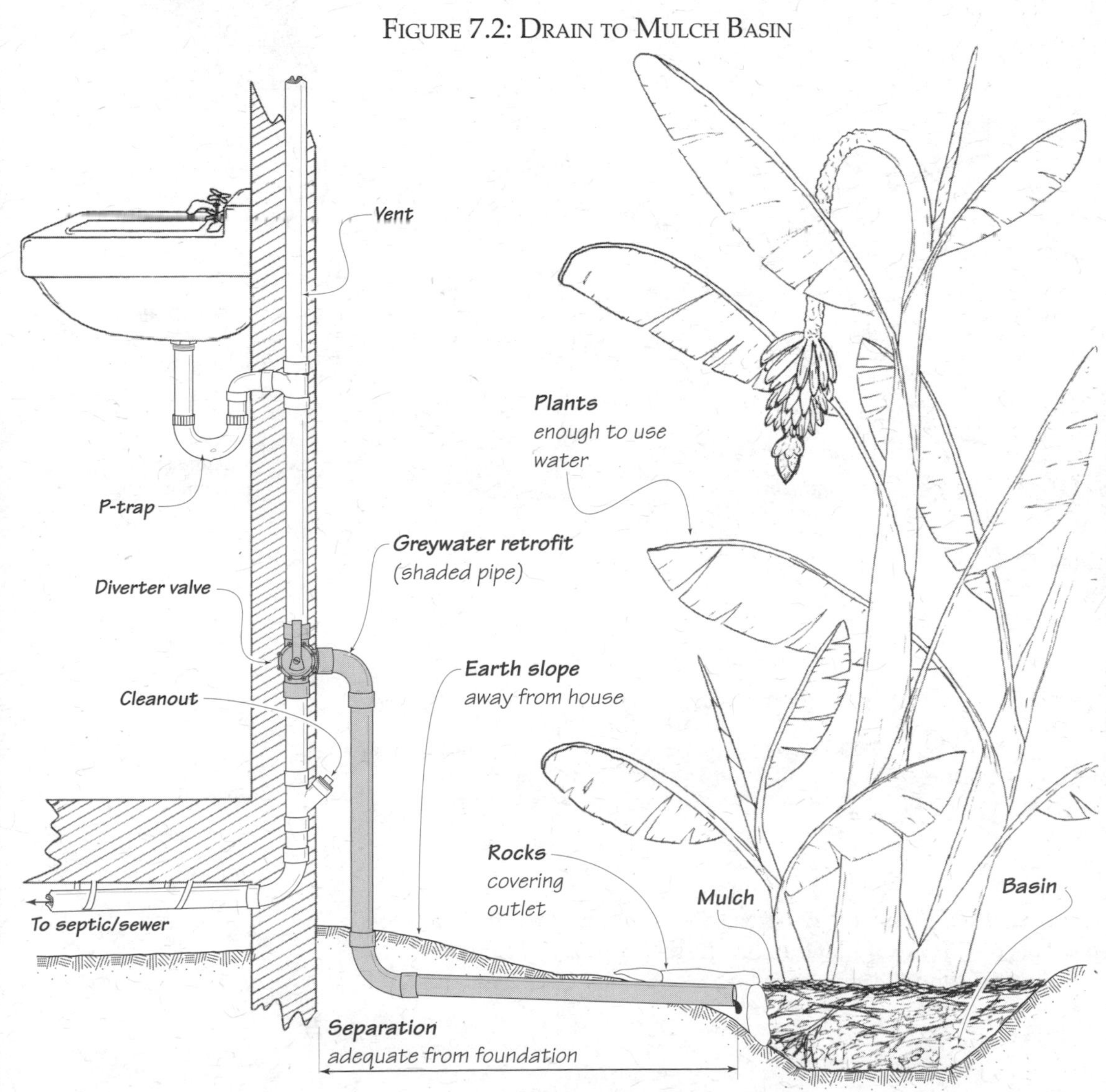

Drain Out Back in Costa Rican rainforest. The water rains into the mini-jungle below, which absorbs and treats it. There isn't even a drainpipe. This system has worked for decades without inter vention or problems.

Skip down

Movable Drain (DR)

An excellent provisional system to use while a new home is being built. As the permanent landscape is installed it can be changed to a Branched Drain or other system. Unfortunately, flexible PVC is an environmental disaster in its production and disposal. Any PVC is bad, but the flexible version has plasticizers, which make it worse. HDPE or 1" garden hose work, but not nearly as well. The application is for an installer or community to have one set of flexible PVC hoses that are passed through many temporary systems.

Flexible PVC is a working tinkertoy set for moving water or wastewater without glue (or leaks) at pressure or vacuum up to 10 psi.

This design is another Art Ludwig original. From the collection plumbing, route a standard ABS waste line at ¼" per foot (2%) slope to the edge of a mulch basin. Run the greywater into this basin, or route it to other downhill basins through press-fit lengths of 1" flexible PVC (spa-flex).[s6]

Flexible PVC is inexpensive, makes nice sweeping curves, and is impossible to kink. The dark brown version looks better snaking around the yard than the common brilliant white variety. At 1" the diameter is a large enough for a tub to drain tolerably quickly, eliminating the need for a surge tank to accommodate slow flow through a narrow hose bottleneck.

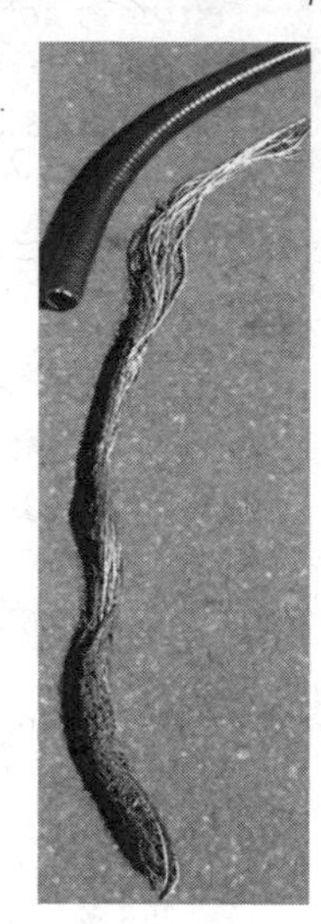

This flexible line wasn't moved for just six months, and an arm-length root plug was the result.

The collection plumbing for this system is tedious and expensive (as gravity collection plumbing always is), but the distribution is simplicity itself. All of the crud that would choke a filtered system flows harmlessly out into the landscape and becomes compost.

There is a bit of a learning curve for laying hose for non-pressurized greywater. Hose lengths should be no longer than needed, and the slope as constant as possible, so air and water can drain freely. The main obstacle is an "airlock," where air trapped in an inverted U-section blocks the flow of water. Right-side-up U's appear to work fine, at least with low solids, high volume greywater, and occasional moving of the hose.

Figure 7.3: Movable Drain

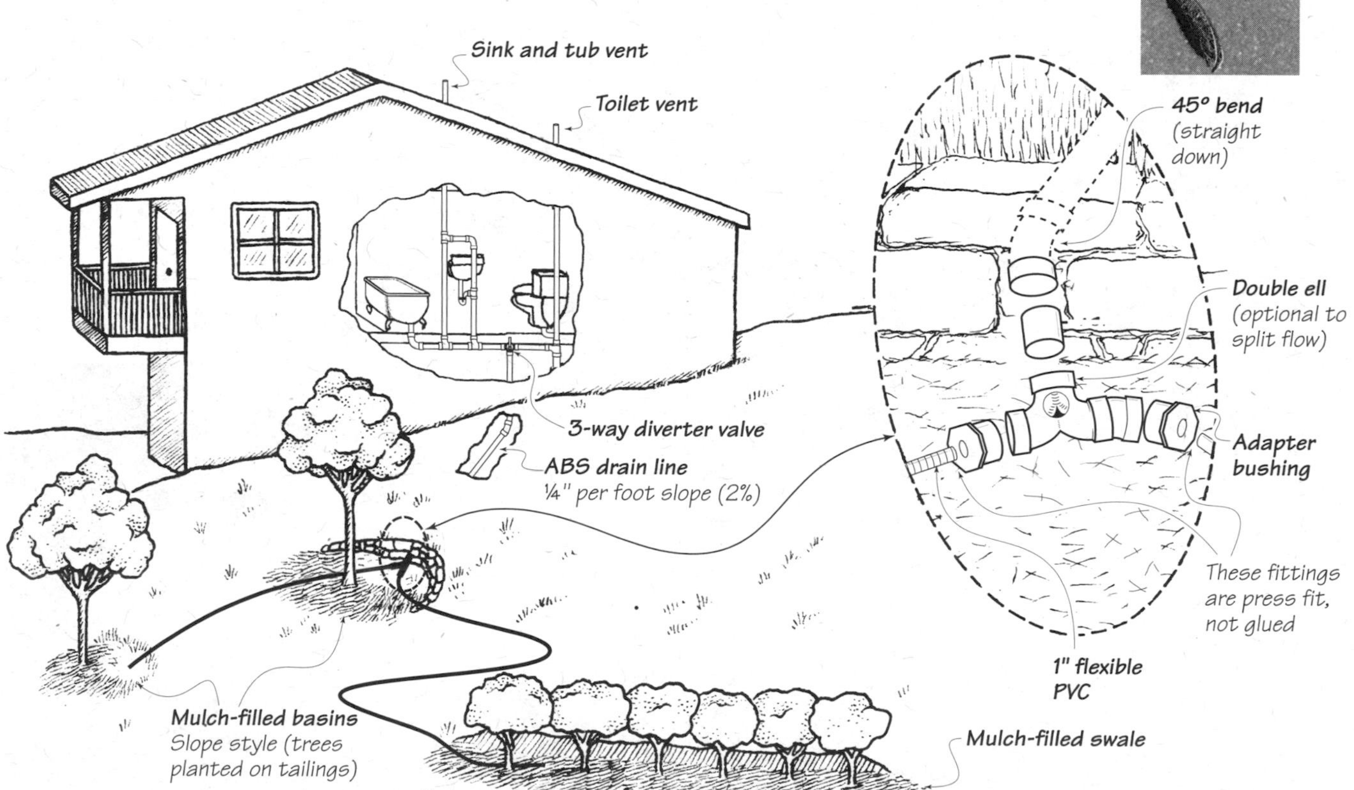

Branched Drain (DR)

The system I recommend most frequently if the area to be irrigated is downslope from the greywater source and the volume is residential scale. It is not simple to build—with lots of greywater and little slope, it is quite challenging—but it is easy to use and maintain. (An Art Ludwig original design.)

Branched Drain systems improve on the Drain Out Back by splitting, containing, and covering the flow. The flow is split by double ells or other flow-splitting fittings in a branching network (right). The output is contained in basins or Subsoil Infiltration Galleys, covered by mulch or soil.

For sites with continuous downhill slope from greywater source to irrigated areas, **Branched Drain systems provide inexpensive, reliable, automated distribution with almost no maintenance.** Branched Drains have no filter, pump, surge tank, or openings smaller than 1" *(2.5 cm)*.

All variations of this system meet legal requirements in Arizona, New Mexico, or Texas (see Appendix G). A friendly inspector can issue a permit for this system under the CPC/UPC. Chapters 9 and 10 explore every aspect of Branched Drain systems in detail.

Figure 7.4: Branched Drain Networks (plan view)

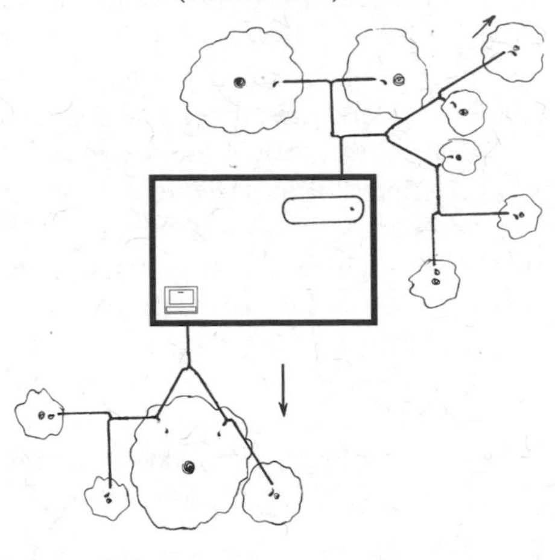

Laundry Drum (CDR)

The classic design for a laundry-only system. If you'll actually keep moving the hose 50–100 times a year, year after year (most people won't), this is a great option for greywater reuse. If you don't, it still works, but as a disposal rather than a reuse system. Plumbing* all *your greywater sources to the drum works too; it is then called a Gravity Drum, which covers D and R, the missing C being the house collection plumbing.*

With this system (Figure 7.5 and photo at right), even distribution is achieved by moving the hose from mulch basin to mulch basin. This deceptively simple approach elegantly solves a host of problems, in exchange for one ongoing task: Someone has to move the hose every couple days. If you're a person who spends time in the garden and will move the hose, seek no further—this is the most reliable way of delivering laundry greywater to your plants and requires virtually no maintenance, decade after decade. The same distribution system could be connected to collection plumbing from the shower, bath, etc.

Figure 7.5: Laundry Drum

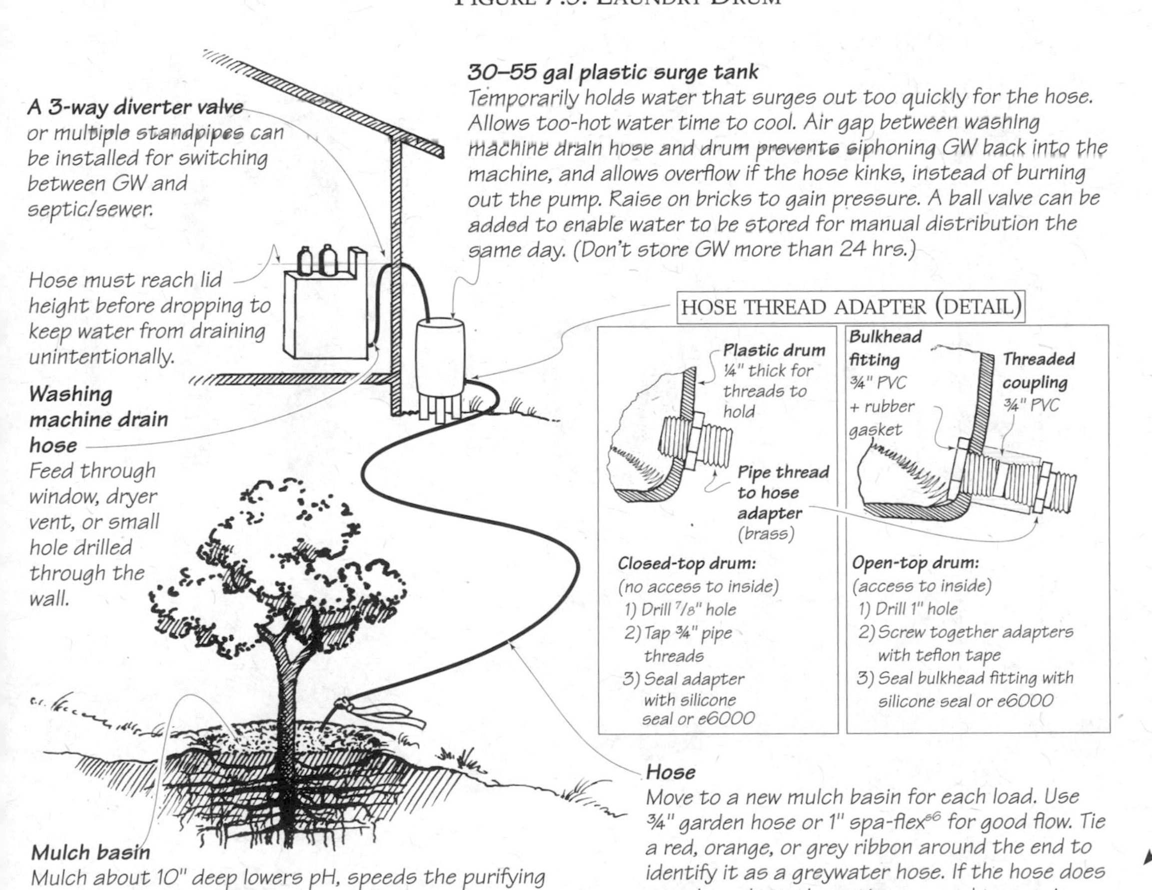

A 3-way diverter valve or multiple standpipes can be installed for switching between GW and septic/sewer.
30–55 gal plastic surge tank
Temporarily holds water that surges out too quickly for the hose. Allows too-hot water time to cool. Air gap between washing machine drain hose and drum prevents siphoning GW back into the machine, and allows overflow if the hose kinks, instead of burning out the pump. Raise on bricks to gain pressure. A ball valve can be added to enable water to be stored for manual distribution the same day. (Don't store GW more than 24 hrs.)
Hose must reach lid height before dropping to keep water from draining unintentionally.
Washing machine drain hose
Feed through window, dryer vent, or small hole drilled through the wall.
HOSE THREAD ADAPTER (DETAIL)
Plastic drum ¼" thick for threads to hold
Pipe thread to hose adapter (brass)
Bulkhead fitting ¾" PVC + rubber gasket
Threaded coupling ¾" PVC
Closed-top drum:
(no access to inside)
1) Drill ⅞" hole
2) Tap ¾" pipe threads
3) Seal adapter with silicone seal or e6000
Open-top drum:
(access to inside)
1) Drill 1" hole
2) Screw together adapters with teflon tape
3) Seal bulkhead fitting with silicone seal or e6000
Mulch basin
Mulch about 10" deep lowers pH, speeds the purifying action of soil microorganisms, and keeps greywater from lingering on the surface. A basin 10" deep keeps water where it's needed and prevents greywater from running off.
Hose
Move to a new mulch basin for each load. Use ¾" garden hose or 1" spa-flex[e6] for good flow. Tie a red, orange, or grey ribbon around the end to identify it as a greywater hose. If the hose does not slope down the entire way and traps air, the next load may not drain without first giving the hose a whip to break the airlock.

A Laundry Drum supplied by a washer drain hose run through a 1" hole in the wall. Note this drum also catches rainwater, which runs through a screen mesh and out through the same exit hose to flush salts from the mulch basins.

¾" garden hose to fruit trees

*May be succeeded by the Laundry to Landscape system, next page, if the new design works longterm.

Laundry to Landscape System (CDR)

A new (2008), unproven Art Ludwig refinement of an old idea. If Laundry to Landscape systems keep working as well as they have so far, most homes should have one. It is a cost-effective retrofit system, and renter-friendly. It can irrigate areas level with or slightly uphill from the washer. This is the system I recommend most often, usually in combination with a Branched Drain or Green Septic system.

People commonly attach garden hoses to their washers. There are several issues with this practice (listed on p. 106), which the Laundry to Landscape system addresses (Figure 7.6).

It uses 1" polyethylene (the most "eco" plastic) to accommodate the rush of water from the washer, without a surge tank or stressing the pump. Thus, **the washer itself pumps water a large distance horizontally, or a short distance vertically, to multiple outlets** (six to 18 of 'em), ***without moving a hose.***

The diversion occurs upstream of the standpipe (the drain the washer outlet hooks into). You don't have to mess with the house drain plumbing. (This is a point we're stressing in our campaign to have California allow their installation without a permit or an inspection.)

The current (March 2009) state of the art design follows...

We're compiling evolving best practices for Laundry to Landscape systems, including a list of parts and where to get them, and free "open source franchise" information to run a business installing them, at oasisdesign.net/greywater/laundry. If you have design tips or photos to share, please email us or post to our greywater forum.

Washer Pump Performance and Distribution Plumbing Limitations

Laundry to Landscape systems use the **washing machine pump (A,** in figure 7.6**)** to distribute the water. Without stressing the pump you can irrigate any distance downhill, or pump up to an elevation 2' below the top of the washer 100' away (100' of horizontal 1" tubing offers the same resistance as 20" of vertical rise). The resistance the pump has to overcome

FIGURE 7.6: LAUNDRY TO LANDSCAPE SYSTEM

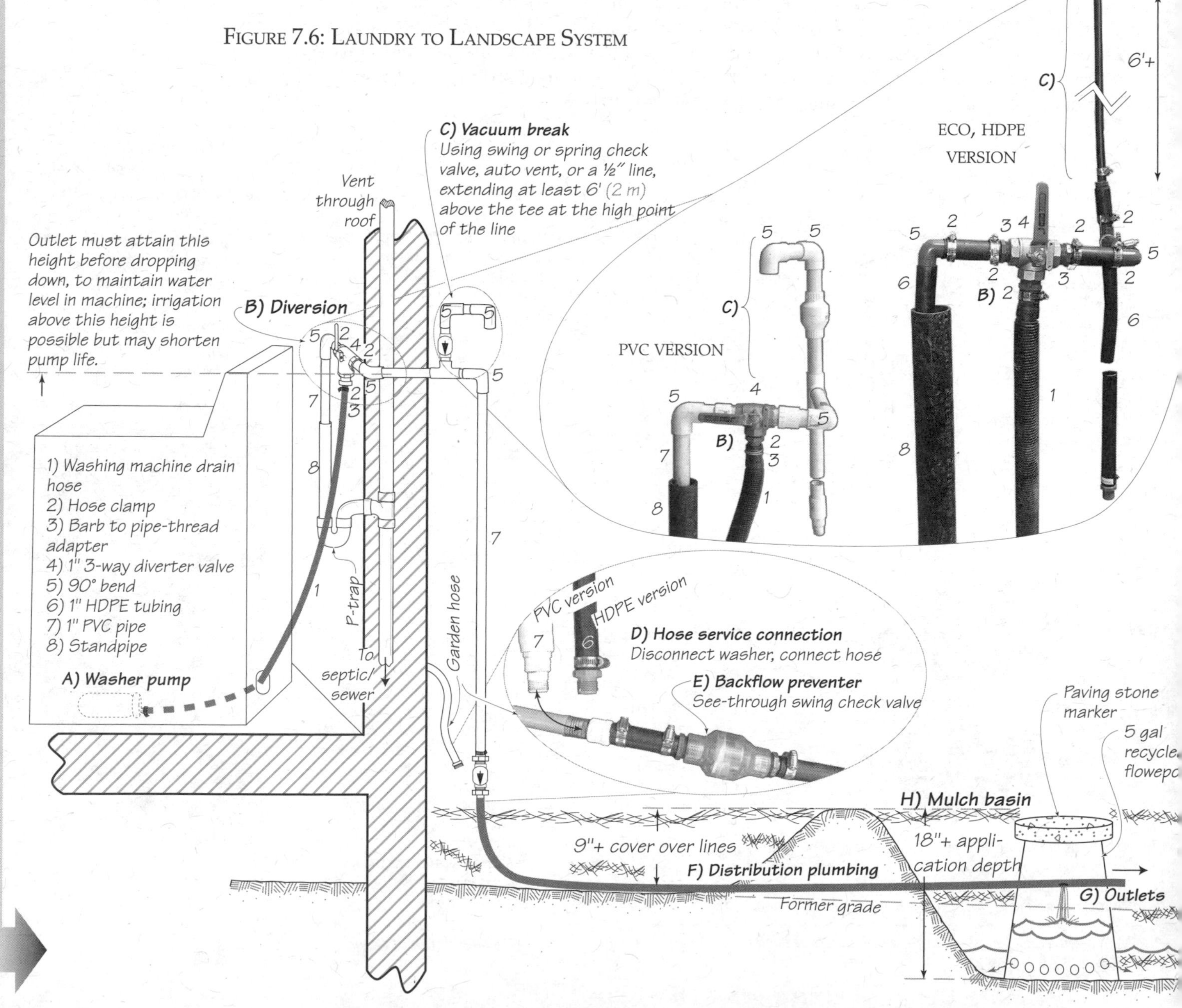

should ideally be about the same as in a standard installation, where the hose discharges at the height of the top of the machine. For example, if a washing machine empties through 100′ of 1″ pipe that ends 18″ lower than its lid, the effective resistance is the same as if it discharged 2″ above its lid. At considerable risk to the pump, I've seen people irrigate up to 6′ above the top of the washer. [m] The variables that affect pump life are:

- **Pump model**—Higher-quality pumps perform better. Unfortunately, every washer pump is different. The way to determine if the pump is adequate is to try it and see if it burns up.
- **Height differential**—The less rise, the easier on the pump. I wouldn't go more than 6′ up.
- **Pipe flow resistance**—The pipe should not be less than 1″diameter, and should not kink.

Leveling a Laundry to Landscape line buried 4" deep in a hard-freeze climate. This line ran fine all winter, even though the plumbing inside the house froze!

Diversion

The **laundry diverter valve (B)** is usually mounted on the wall behind the washer, or where it is easily visible and convenient to turn. It should be solidly screwed to the wall using copper pipe brackets or plumber's tape, so that it does not wiggle when the handle is torqued. One side of the valve diverts water into the standpipe through an air gap, the other through the wall or window to the outside. The greywater destinations should be clearly labeled, eg., "ocean" and "citrus trees."

Vacuum Breaker

If the first outlet is lower than the level of water in the washer, a **vacuum breaker (C)** is advised to keep the drain line from continuously siphoning water out of the machine as it tries to refill itself (not a problem with every machine or load, but...easier to just add it). The loose fit of the washing machine drain hose into the standpipe in conventional plumbing creates an air gap, which serves as a vacuum breaker. The vacuum breaker *must connect to the main line at its high point to be effective*. This is typically close to the washer, just outside the house. If the line must dip down before leaving the house, it could siphon even with a vacuum breaker outside. In this case, mount the vacuum breaker inside. To avoid the possibility of spillage indoors, you can route a ½″ tube from the top of the vacuum breaker outlet back into the septic/sewer drain standpipe.

If you're irrigating uphill and the first outlet is above the top of the washer, it will serve as the vacuum breaker.

Automatic Bypass (freezing climate only)

If the line could freeze, you *must* have an automatic bypass. This is a pipe through which the water rises and harmlessly overflows when the line clogs with ice instead of burning out the pump or flooding the house. The ideal bypass makes an audible splash so you know what's up. The tall vent vacuum breaker option shown as a ½″ line at the right of Figure 7.6 could double as an automatic bypass if it were made of 1″ tubing. You'd certainly notice when greywater fountained all over the side of your house. Perhaps one of you from a freezing climate will come up with a better design with less drastic notification of frozen lines. If the distribution lines slope downhill continuously they probably won't freeze (photo).

Hose Service Connection

A **hose service connection (D)** makes tuning the outlets easier (you won't have to keep doing load after load of laundry to check and tune outlet flows). It is also good for blowing out lint, if needed (or blowing the system apart if you pressurize it more than the 20 psi *(140 kpa)* that this type of plumbing is designed for). The hose service connection must be properly installed so there is no chance of greywater backflowing into the freshwater lines. The layers of protection against this are: 1) to connect the hose, the washer must be disconnected; 2) the swing check valve (below); 3) a backflow prevention device at the hose bibb.

To tune the outlets perfectly, check the flow from the washer by timing how long it takes to fill a bucket. Then adjust the hose to the same flow.

Backflow Prevention Valve

If the drain line runs (or can be lifted) above the height of the top of the washer, a **swing check valve (E)** should be included as close as possible to the washer, to keep water in the line from rushing back into the machine when it shuts off. Get a clear one with 1″ pipe thread—a clean installation and you can watch what is happening inside.

If you have a hose service connection, a swing check valve adds backflow protection.

[m] ***Metric:*** *... can irrigate up to 50 cm below top of washer 30 m away (30 m of 2.5 cm hose = 50 cm vertical). 30 m of 2.5 cm pipe ending 45 cm lower than lid = resistance of 5 cm above lid. At risk to pump, up to 2 m vertical rise is possible.*

Distribution Plumbing

To get the pressurized greywater to plants, 1″ polyethylene tubing is the preferred **distribution plumbing (F),** ideally the kind with a purple stripe to indicate non-potable water. Smaller tubing gives too much resistance. Bigger tubing traps more septic water and crud and is a waste of plastic. PVC pipe destroys the environment, and is ugly.

You can run a single or multi-trunk line, with or without valves or branches. Branches can be 1″, ¾″, or ½″. With lots of greywater and/or low perk soil, use two or more valved zones (photo). All the plumbing can be under 9″ of mulch for a California Plumbing Code appendix G legal system (Figure 7.6), otherwise it can just go on top of the mulch.

It is best for freezing, smells, and the pump if the line slopes downhill continuously. Second best: a U-shaped line with an outlet at the low point to drain the U.

However, because the line is pressurized, it can dip up and down. The consequence is some trapped water in the line between uses. Unless the line might freeze, this is acceptable. The water can go septic if it sits for more than a few days. However, the quantity in even a long run of 1″ pipe is so small that any objectionable smell is only detectable for the first moment of discharge. In an installation that includes both some rise and a long horizontal run, the quantity of trapped water is minimized by sending the pipe up to the maximum height as quickly as possible, then running the pipe down from there. This way, most of the run drains dry after each use. This same geometry works to get the water up from a basement washer to the yard in a freezing climate, as this places the part of the line that holds standing water inside the thermal envelope of the house (you could also add a surge tank and effluent pump).

Zone valves in valve box made from a recycled plastic drum.

OUTLET OPTIONS

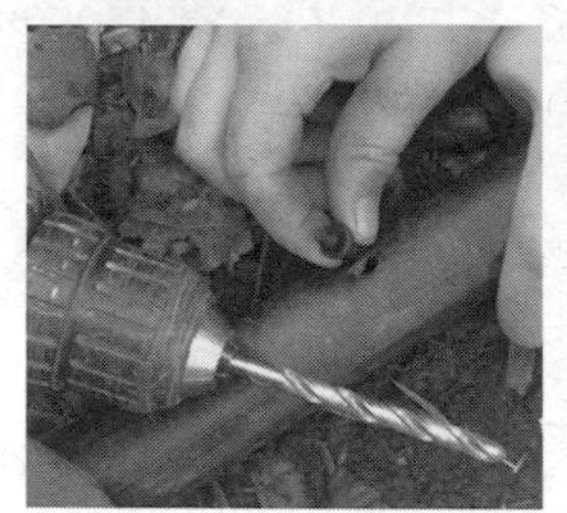

Hole drilled in tubing as emitter. Seems like cheating, it's so easy, but they're working great so far. Put them on the bottom of the line so greywater isn't geysering everywhere. A 7/32″ (5.5 mm) hole is the right size to be able to plug them with drip "goof plugs" if you change your mind about the location.

1″ x ½″ tee outlet.

½″ ball valve outlet tee (have to shorten barb).

Outlets

The capacity of all the **outlets (G)** should be enough that the pump is not strained trying to push too much water through too small or too few holes. On the other hand, too many or too large holes will result in pressure loss that may leave some outlets high and dry. The total cross-sectional area of all the outlets in a zone should be 1–2 in^2 (the Laundry to Landscape Calculator at oasisdesign.net/greywater/laundry can be used to find the total cross-sectional area from a variety of outlets).

Note that outlet flow in this pressurized system, unlike a gravity flow system, depends on the height, the size, and the number of outlets, as well as the length and diameter of the tubing. You can tune the flow by making the outlets different sizes, or adjusting the outlet ball valves (photo below). If you are irrigating uphill, the first outlet will get way more water than the last outlet. To avoid this, run a solid line to the high point, then do a U-turn and put all the outlets in the downhill run.

Receiving Landscape

Mulch basins (H) are the way to go. With the dimensions indicated in the drawing, they can be legal under the CPC. Otherwise, the hose can go on the surface and the outlets can be directed straight down into the mulch.

Not every installation requires a vacuum breaker, check valve, hose service connection. But, if you include the applicable components, the chance of having trouble with your system is much smaller, and including all of them won't hurt.

Outlet shield in mulch basin (5 gal pot).

Garden Hose through the Bathroom (CDR)

Commonly used as a temporary drought-emergency measure.

In the bath or shower, block the drain with a flat rubber drain stopper, then siphon the water through a hose from the bottom of the tub or shower stall into the garden after you bathe or as you shower. Use the warm-up water (the cold water that comes out of the tap at first) to start a siphon. The "pro" way to start a siphon is to fill the hose completely with water (using a shutoff valve on the end, if necessary), then open the outlet while the inlet is underwater. A sump pump can be used as well (do not siphon greywater by mouth—duh!).

Dishpan Dump/Bucketing (CDR)

A good choice when water is extremely tight and time is abundant, in a drought emergency, or if the water supply system goes down and you need to make the most of greywater to keep plants alive.

The Dishpan Dump is the most time-honored of all greywater systems: When the dishpan water gets dirty, dump it in a mulch basin next to plants.

No system can equal the efficiency attainable by Bucketing, even at a thousand times the cost. It can be the perfect greywater solution for those rare applications where water is truly precious, gardens need irrigation, and people have energy to haul buckets around daily. (A refined variation of Bucketing I call the Multi-Mode Greywater Tank is described in Real World Example #3.)

Mulch Basins (R)

The simplest, best receiving system for greywater.

This is what receives the greywater from all the systems mentioned up to this point. Mulch basins are discussed at length in Mulch Basin Design, Chapter 5.

Greywater Furrow Irrigation (CDR)

An excellent system for non-industrialized countries.

This is a greywater system reduced to its simplest elements: soil and plants. It consists of narrow channels formed in the soil that direct greywater to plants. Developed for a village context, Greywater Furrows can be owner-built with a shovel, hoe, or rake. If you are interested in serious simplicity, this system is worth checking out—details can be found in Appendix D. It is best adapted to non-industrialized world conditions where exposed greywater flows are the norm and mulch is unpopular.

Before: *Greywater puddles in public street in Mexican village.*

After: *Village women have made Greywater Furrows (arrows) that channel laundry water and bathwater to several newly planted fruit trees.*

GREYWATER BASICS ❖ GREYWATER BASICS ❖ GREYWATER BASICS ❖ GREYWATER BASICS

Continue p. 74

Chapter 8: More Complex Greywater Systems

The simple systems in Chapter 7 may not be an option if your plants are uphill from your greywater source, your greywater flow is high, it makes sense to reuse your blackwater as well as greywater, automated high efficiency distribution is desired, you need CPC/UPC compliance, or some other reason. If this is the case, check out the systems that follow...

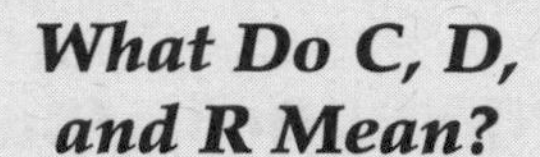

C: Collection plumbing
D: Distribution plumbing
R: Receiving landscape

These are the parts that together make a complete greywater system.

Drum with Effluent Pump (D)

The cheapest uphill distribution system. The pump quality and design details greatly impact its reliability. Virtually every one of these that includes a filter and a cheap pump is eventually abandoned due to maintenance hassles. If you choose this system and want it to last, go for an effluent pump capable of passing ¾" solids, and forget the filter.

This system consists of (surprise!) a plastic drum for a surge tank, a filter (optional, and emphatically *not* recommended), and a pump (Figure 8.1). You'll also need some means of distributing and receiving the water in the landscape, such as Mulch Basins or Mini-Leachfields.

If your garden is downhill from the drum, which may be up to 5' higher than the washing machine, a Gravity Drum (that is, drum *without* pump or filter) that costs $20–$100 is best (see previous chapter). Otherwise, I recommend you bite the bullet and purchase an effluent pump. These are not cheap (see Appendix E).

If the irrigated area is uphill, you'll want a backflow prevention valve.

You can distribute the water in the landscape via a movable hose (most people will quickly tire of moving it) or a branched network of permanent HDPE black drip irrigation tubing. This could consist of main lines of ¾–1" diameter, with branches of ½" to Mulch Basins or Mini-Leachfields. For "emitters," use open-ended ½" drip tubing. Or, follow Larry Farwell's idea of using inexpensive ½" plastic inline drip ball valves to reduce the aperture size at the outlets, and/or to direct the water among different zones; these may clog, but you can just open the valves wide once in a while to free them.

FIGURE 8.1: DRUM WITH EFFLUENT PUMP

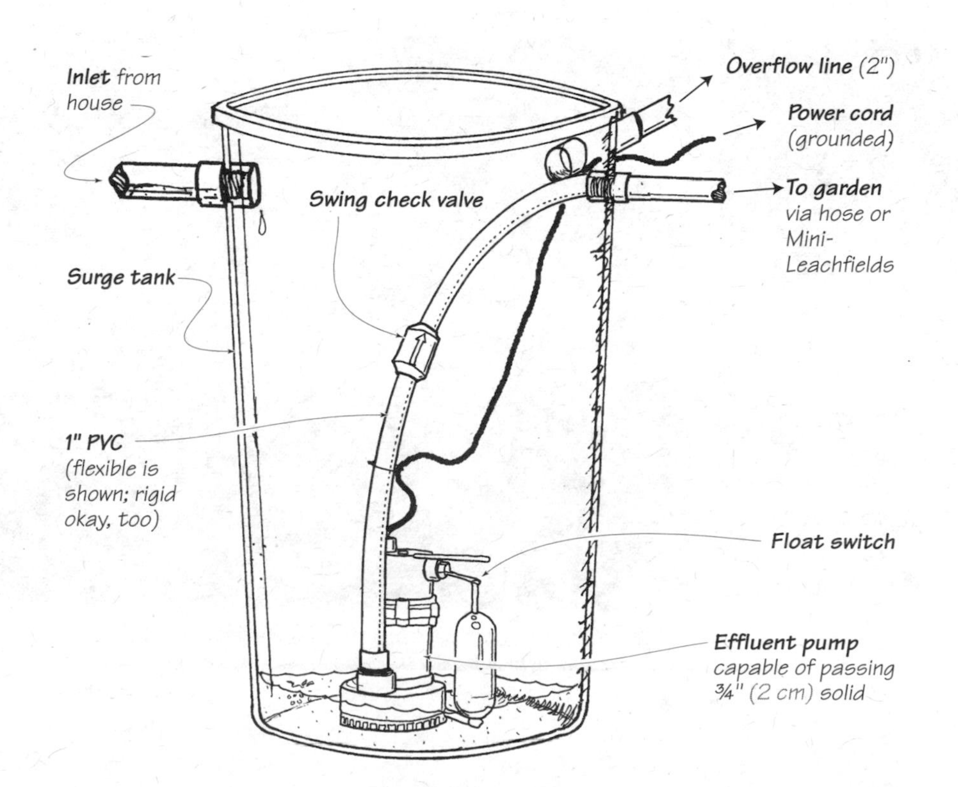

Effluent pump. The expensive ones are well made. The $500 Zabel pump pictured supposedly has a service life of 27 years average, for pumping septic tank effluent. $100 pumps require filtering and rarely last 5 years.

Caution: Don't use drip irrigation emitters, as they will clog on the first use.

Note that the flow won't split predictably as it does when greywatering at atmospheric pressure through a Branched Drain network. Instead, the flow has to be fine-tuned so that outlets at different heights and at the end of different lengths of tubing emit an appropriate amount of water.

To receive the effluent in the landscape, you can use Mulch Basins, Mini-Leachfields, or adequately sized Subsoil Infiltration Galleys.

Drum with Pump and Mesh Filter is pretty much a discredited technology at this point due to longevity problems. It may make sense as a temporary system. A few percent of them have lasted more than a few years, and they remain so popular that against my better judgement I'm including this brief description:

The surge tank is the same, but the pump is a cheap sump pump (see Appendix E). To prevent clogging and hair from wrapping around the pump rotor and killing it right off, tie pantyhose to the inlet as a filter. Many people discard the inexpensive pantyhose rather than clean them. The filter bags from airless paint sprayers work as reusable filters. This system costs $100–$400.[s12] Design it so that when the cheap pump fails or you tire of filter cleaning, you can easily upgrade to the effluent pump mentioned above.

Mesh filtering allows reliable greywater distribution via the irrigation tubing system described above.

Don't use drip irrigation emitters with mesh-filtered greywater. It will work for a while, but then they will clog—this has been firmly established. If you want to use drip irrigation, you'll need an auto-backwashing pool sand filter or media filtration—systems discussed toward the end of this chapter.

Figure 8.2: Mesh Filter Bag

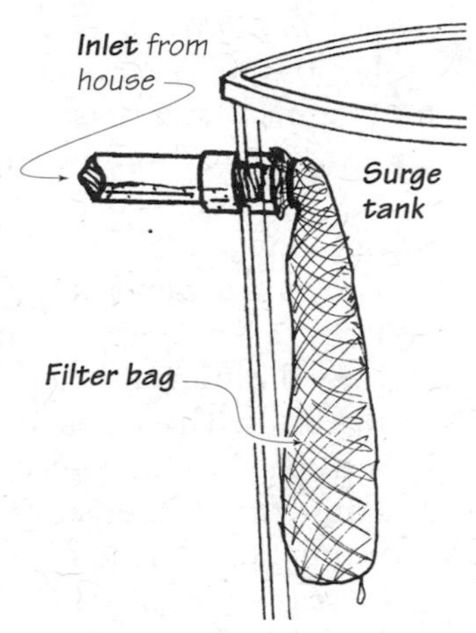

Mini-Leachfields (R)

A reasonable compromise when subsurface irrigation is needed in a legal system.

Mini-Leachfields offer somewhat even subsurface irrigation at a reasonable price (Figure 8.3). The original version is by Larry Farwell, the man largely responsible for the legalization of greywater in California. It uses a tee, a short length of ½" tubing that protrudes through the bottom of an inverted flowerpot, and another tee on the end to keep it

Figure 8.3: Mini-Leachfield

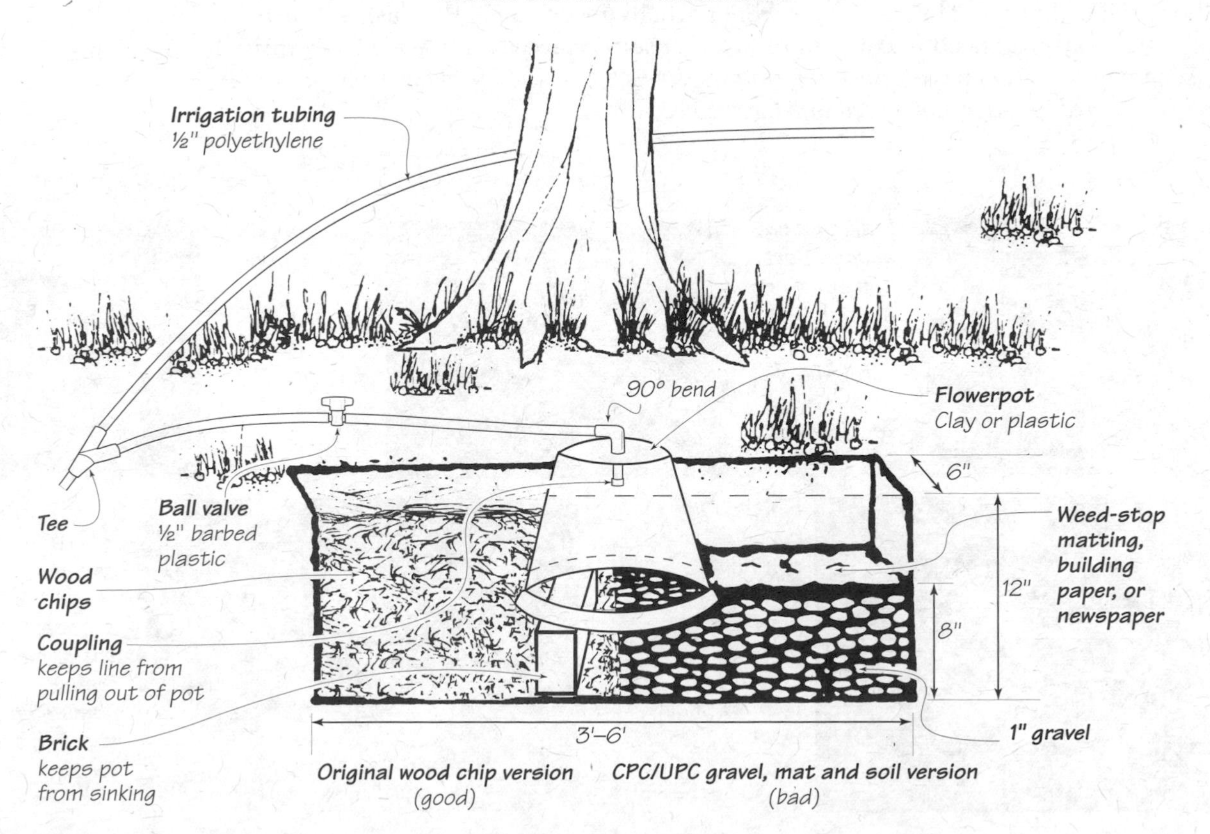

from pulling out of the pot, which discharges into a mulch-filled pit or basin.

The Mini-Leachfields described in California greywater law are completely buried, with gravel under the emitters. For people who, like me, are concerned about checking for clogs and unenthusiastic about shoveling gravel into the garden, Mini-Leachfields could be filled to the top with wood chips in place of gravel, filter fabric, and soil.

According to Larry, two gravity-fed Mini-Leachfields can be supplied simultaneously through ½" dripline (make sure that they are at equal elevation). If you have a pump, you can supply six outlets.

Caution: I think the original version makes sense, but wouldn't use the overkill version described in the CPC/UPC anywhere. You'd be better off adding to your septic system.

Subsoil Infiltration Galleys (R)

Simplest method if subsoil distribution is required. Sizing is important, as is access for service in the distant future. I often use these for kitchen sink effluent even if not legally required.

Infiltration Galleys are open spaces under the soil into which you can pour effluent for treatment and reuse or disposal. They provide surge capacity equal to their volume and wastewater infiltration area equal to their open floor area (and open wall area in the case of infiltrators, described below). They are so safe they've been used successfully for raw blackwater.[24]

Greywater can be supplied to Infiltration Galleys via a gravity Branched Drain network, a Drum with Effluent Pump, or a Laundry to Landscape system. They do not require filtration, although an impermeable biomat may form in the absence of active earthworms. They must be supplied by multiple outlets, or be perfectly level and dosed in large batches for even distribution (see Dosing Siphon, Chapter 4). The optimum size dose (greywater surge) would fill the entire trench 1½" deep. Dosing prevents an accumulation of solids at the inlet. Otherwise, progressive infiltration failure can occur as successive small areas are overwhelmed and clogged with solids.

Consider Infiltration Galleys' ability to support the weight of soil and traffic, ease of inspection/service access, resistance to clogging by roots, and environmental impact/cost of materials. The various means of making these spaces include infiltrators, box troughs, and half-pipes or drums.

Infiltrators are gravelless infiltration galleys now used for the majority of new septic leachfields in the US. Their form resists the weight of traffic, and they are designed to emit water not only through the open bottom but also halfway up the louvered sides, without soil leaking in. They are made from partially recycled HDPE plastic, a fairly benign material. You need to cut your own larger service access and make it a soil-tight lid (see images on this page). *Note: Infiltrator Systems*[s13] *has redesigned these with narrower lower slots to reduce migration of soil into the chambers.*

INFILTRATORS

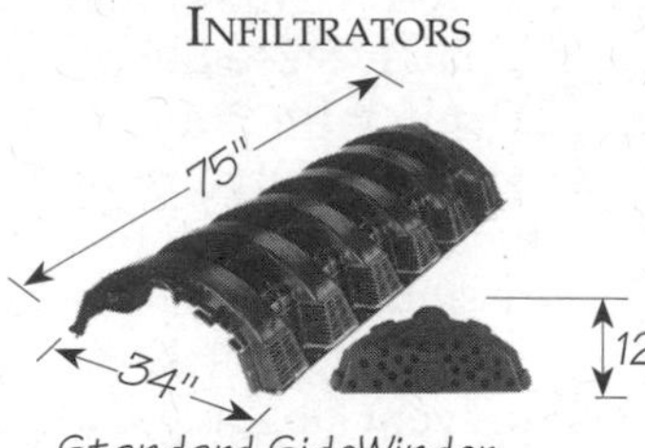

Standard SideWinder
17 ft^2, 76 gal
1.7 m^2, 0.29 m^3

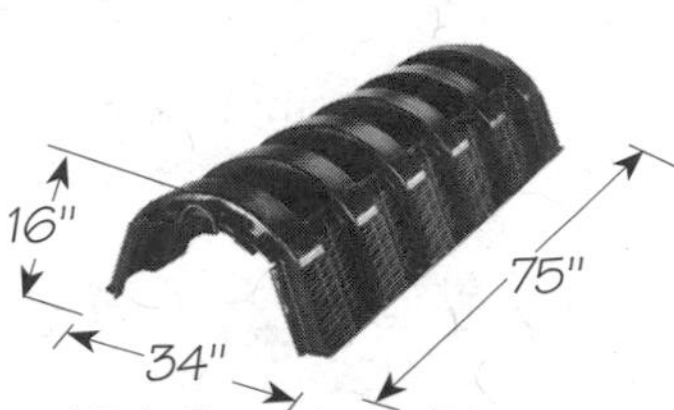

High Capacity Infiltrator
17 ft^2, 103 gal
1.7 m^2, 0.39 m^3

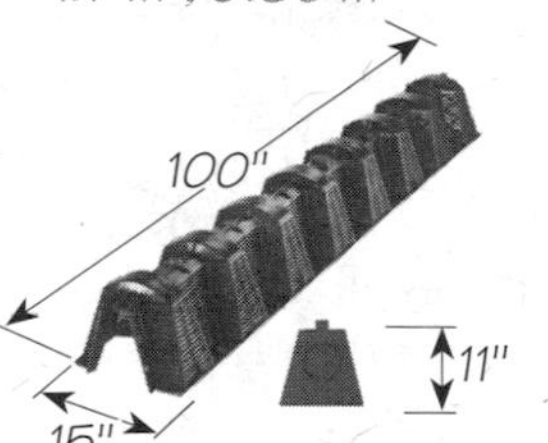

Equalizer 24 Chamber
10 ft^2, 34 gal
1 m^2, 0.13 m^3

Making a lid, access port: Cut out a section between structural ribs with a sawsall or keyhole saw.

Fill around it with concrete. This lid includes a smaller inspection port.

Finished access lids, one per infiltrator. They will have pavers over them on the surface to mark their location.

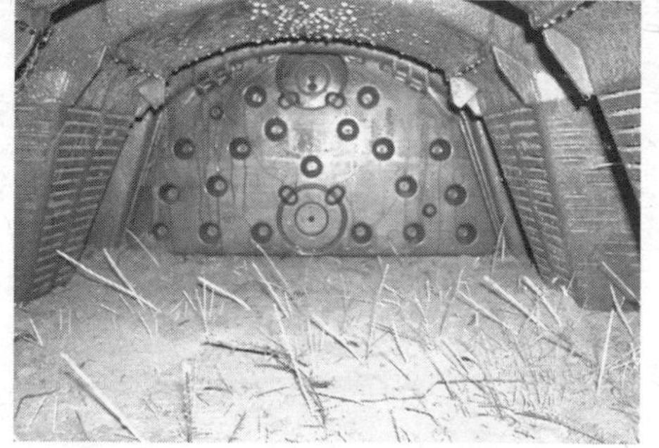

Photo taken by holding the camera down inside the access. Note the bamboo feeder roots coming up through the earth floor.

Box troughs are narrow, rectangular chambers made of flat material such as wood (Figure 8.4).[s14] They can water a planter in a greywater greenhouse, for example (photo at right). The advantage of box troughs is completely open access for inspection/service. A Branched Drain network for distributing the dose internally could possibly be concealed inside the box trough. (Contact us for the latest on this variation, which is promising but untried as yet.[s9])

Box troughs water a greywater greenhouse in Maryland.

FIGURE 8.4: BOX TROUGH

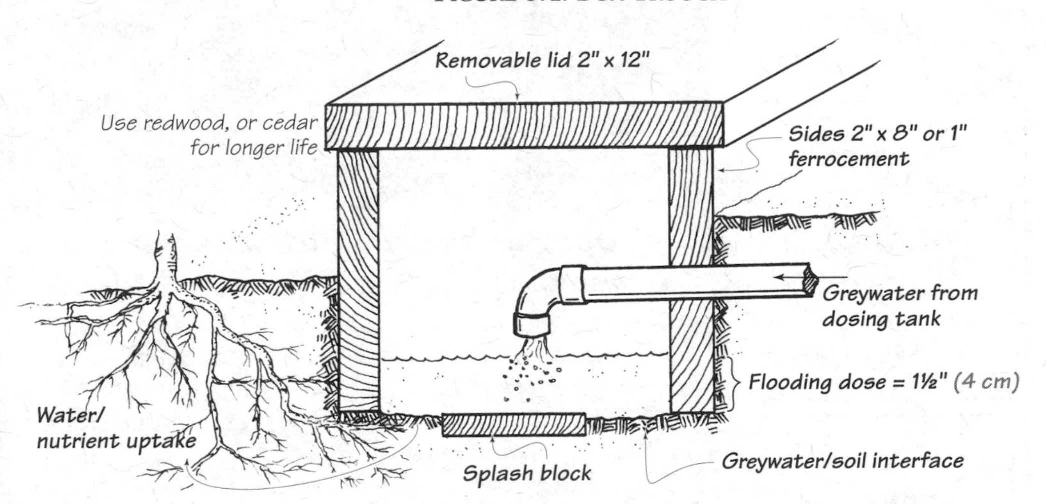

FIGURE 8.5: HALF-PIPE LEACHING CHAMBER

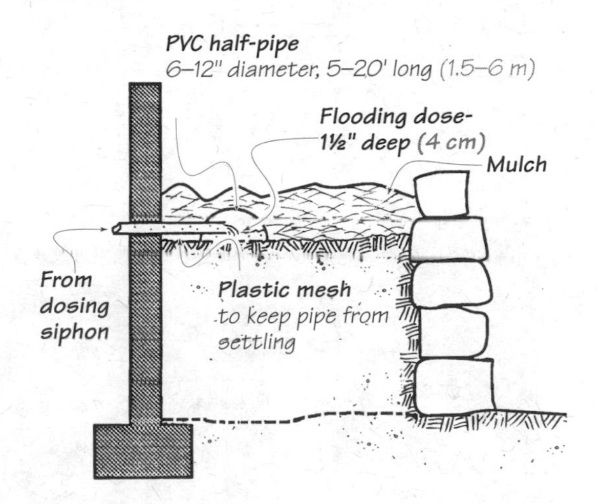

Half pipes or **half drums** can be made from 8–12" *(20–30 cm)* pipes or 15–55 gal *(50–200 L)* plastic drums cut in half lengthwise (Figure 8.5). Rest the half-pipes on 1" plastic mesh or bricks to keep them from sinking into the soil, then cover with soil or mulch. If covered with mulch, service will obviously be easier. The ends should be capped or sealed so mulch doesn't leak into the chamber.

Solar Greywater Greenhouse (R)

The way to go in really cold climates. Successful with box troughs supplied by a Drum with Effluent Pump.

No finer synergy can be obtained between the different systems of a home than with an attached Solar Greywater Greenhouse. Not only does it provide an optimal microclimate for year-round greywater reuse/treatment in the harshest of climates, it also helps heat your house, grows food, and frequently ends up becoming the most popular hangout all winter (see sidebar next page).[25,26]

Compared to outdoor irrigation, a greenhouse's water need is relatively constant throughout the year, and generally, so is greywater supply. Greywater use/treatment may be feasible in a greenhouse even when the ground outside is frozen solid. A solar greenhouse can dramatically reduce space heating requirements.

According to Carl Lindstrom, several greywater irrigated greenhouses in New England have been in operation since the 1970s. One greenhouse uses a clear fish pool with a waterfall and biofilter plates on the bottom as the final treatment. Deep soil beds store heat from the sun and from the greywater itself. This greenhouse was the highest cold-weather producer of greens in the US. These soil beds operate odorlessly and can provide salad greens throughout the long New England winters[s15] (photos next page, front cover).

FIGURE 8.6: ATTACHED SOLAR GREYWATER GREENHOUSE

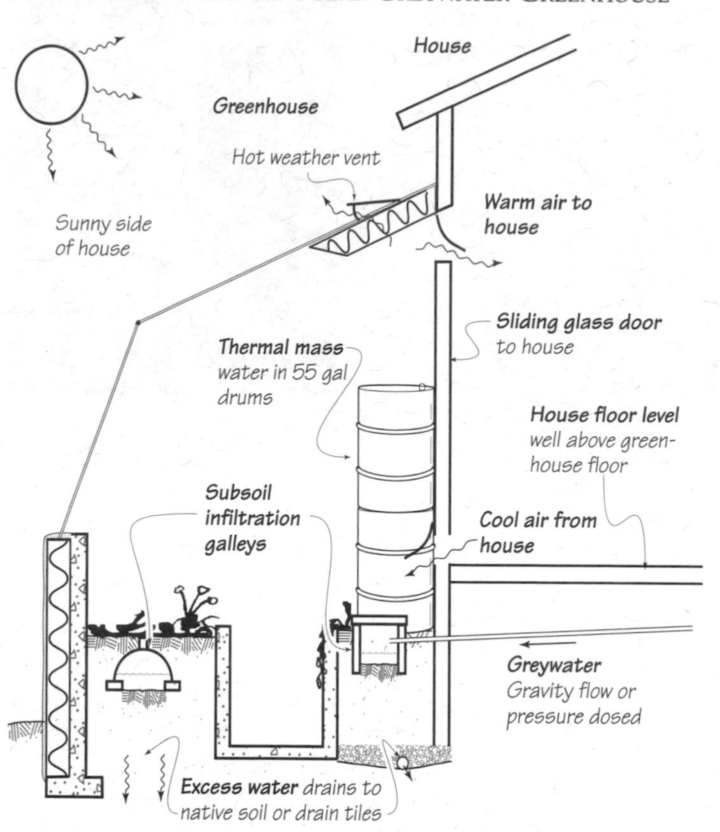

CLIVUS MULTRUM

Greenhouse with Clivus infiltration beds in Massachusetts.

THE FOOD AND HEAT PRODUCING SOLAR GREENHOUSE

Doug and Sara Balcomb's Solar Greenhouse in Santa Fe, New Mexico

The heating for the house is 93% solar: 73% passive and 20% active solar from rock beds under the greenhouse. The owners estimate that this design would fulfill over 50% of heating needs for a home with good solar exposure anywhere in the US.

The two-story, well-insulated house has a 14" adobe wall between house and greenhouse. The home generally stays between 71° and 75°F in the summer and between 65° and 71°F in the winter (22–24°C and 18–22°C, respectively). Because the floor and walls are warm, the owners found themselves more comfortable than in a forced-air-heated home. Here's a quote:

"It is important to emphasize the quality of life that a well-designed solar greenhouse affords the owner. Imagine a blustery, snowy winter day...cold and raw. Then imagine what it is like to step directly into a room where roses and petunias are in full bloom, where lemons are ripening on the tree, where the smell of fresh flowers and green plants pervades. Imagine what it is like to sit in a space like that and sunbathe when it is below freezing outside. Finally, and best of all, imagine picking all the fresh vegetables you and your family need for your winter dinners… This 'impossible dream' is a reality for us, one we hope every family in America will share."[25]

THE FOOD AND HEAT PRODUCING SOLAR GREENHOUSE

Green Septic: Tank, Flow Splitters, and Infiltrators (CDR)

May be the best way to go for a legal system in areas with lousy greywater laws. Especially suited for new construction, replacement leachfields, and water reuse from homes with a large amount of space per person, as it eliminates the need for costly dual plumbing. This type of system can recycle blackwater as well as greywater.

Much as the Branched Drain system is a more evolved, refined version of the venerable Drain Out Back, this system is inspired by the "grass is always greener over the septic system" phenomenon. It consists of conventional combined greywater and blackwater collection plumbing, a conventional septic tank, a dipper distribution box (Chapter 9), and a 2" Branched Drain distribution system that leads the effluent to subsoil infiltrators (see Subsoil Infiltration Galleys, this chapter and Chapter 9). With outlets 9–14" *(23–36 cm)* below the surface, the system is sanitary enough for the reuse of septic tank effluent. This is similar to the standard for secondary treated effluent.

A Branched Drain Green Septic system promises economical, low maintenance, sanitary, predictable, even distribution of clarified septic effluent for irrigating and feeding a bunch of trees. The system's capacity, longevity, and ease of service might well be better than those of a conventional septic system.

Caution: An infiltrator system for even distribution and irrigation is a very promising but unproven variation—trials have been limited and short term. It takes decades for this system to prove its reliability. The stakes of experimentation are high, as replacing a leachfield costs a bundle. Think twice before trying this with low perk soil; better to let this system prove itself and have the kinks worked out in high perk soil first. Current information on this new design can be found at oasisdesign.net/greenseptic.

Note: You don't have to have a full-blown septic system to use the filtration-by-settling principle for greywater only. Any tank with more than two days' capacity coupled with any subsurface distribution system should work.

Blackwater Reuse Health Warning

Don't try blackwater reuse unless you really know what you're doing. Gross misuse of greywater is common, and typically there is no health threat because greywater is just not that dangerous in an industrialized world context. This is not the case with septic effluent. Greywater might have 50 coliform bacteria/100 ml, which is not an unusual level for drinking water in the non-industrialized world; septic tank effluent might have a million. Put another way, your blackwater reuse system must be thousands of times more effective to achieve the same level of safety.

FIGURE 8.7: GREEN SEPTIC: TANK, FLOW SPLITTERS, AND INFILTRATORS
(NUTRIENT AND WATER REUSE FROM A NEARLY CONVENTIONAL SEPTIC SYSTEM)

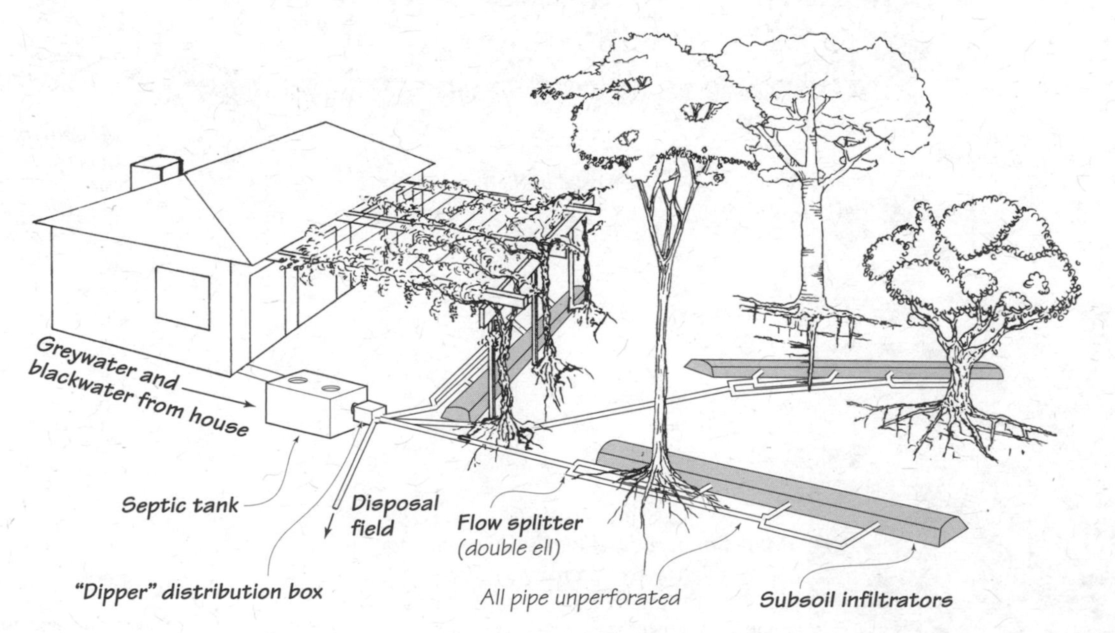

Constructed Wetlands (DR)

These make sense for responsible disposal of greywater in wet climates, and where perk is low and space is tight, and for treatment of larger flows prior to reuse. Constructed Wetlands work for blackwater as well as greywater. They are not generally the best choice in dryland climates, as reuse efficiency is lowered considerably by water loss through wetland plants, especially when evapotranspiration is high and irrigation is needed most.

There are two types of Constructed Wetland: with open water and with water hidden under gravel, with plants coming up out of it (Figure 8.8).[27] Common sense dictates that the hidden-surface type be used for wastewater, at least for the first stage of treatment.

Wetland area is a complicated function of temperature, evapotranspiration, rainfall, influent volume, Biological Oxygen Demand (BOD), and effluent legal standards. A typical size for a wetland in a mild climate is ½–1 ft^2/gal/day, or two or three times this in a cold climate. Wetland depth depends on plant species (see "Water Level" note, Figure 8.8). Although all plant roots can extract nutrients from wastewater, *only* wetland plants function as solar-

FIGURE 8.8: CONSTRUCTED WETLAND

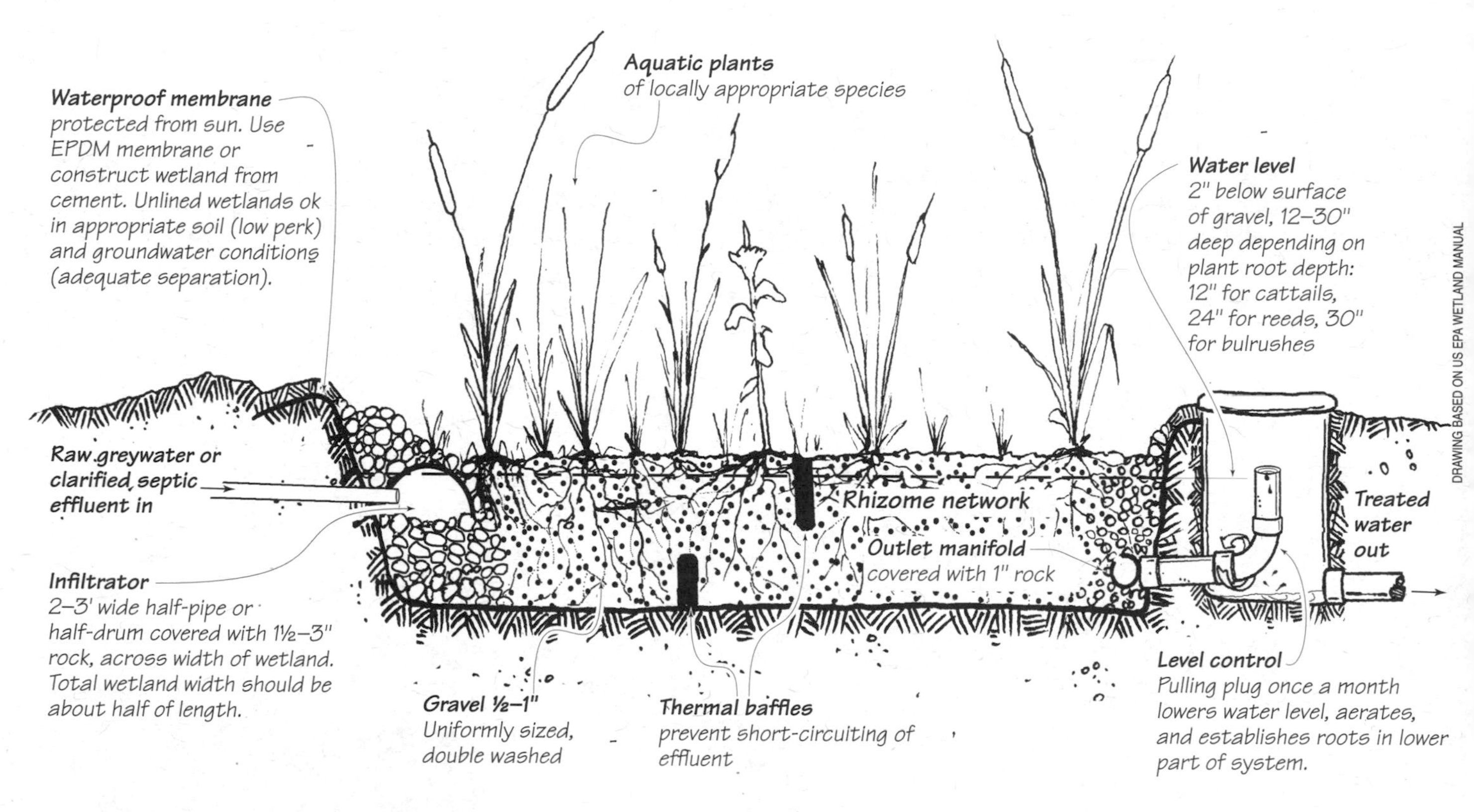
Aquatic plants
of locally appropriate species
Waterproof membrane
protected from sun. Use EPDM membrane or construct wetland from cement. Unlined wetlands ok in appropriate soil (low perk) and groundwater conditions (adequate separation).
Water level
2" below surface of gravel, 12–30" deep depending on plant root depth: 12" for cattails, 24" for reeds, 30" for bulrushes
Raw greywater or clarified septic effluent in
Rhizome network
Outlet manifold
covered with 1" rock
Treated water out
Infiltrator
2–3' wide half-pipe or half-drum covered with 1½–3" rock, across width of wetland. Total wetland width should be about half of length.
Gravel ½–1"
Uniformly sized, double washed
Thermal baffles
prevent short-circuiting of effluent
Level control
Pulling plug once a month lowers water level, aerates, and establishes roots in lower part of system.
DRAWING BASED ON US EPA WETLAND MANUAL

6' (2 m)

Mini-Constructed Wetland in Mexico *turns blackwater from two toilets into irrigation water. The plants are, from left to right: papyrus (taller than a person), canna lilies, and, on the tiny open-surface portion, water hyacinths.*

THANKS TO NATURAL SYSTEMS INTERNATIONAL, LLC AND TAD MONTGOMERY FOR HELP WITH THE WETLAND SECTION.

powered pumps that push oxygen out through the roots. This oxygen keeps Constructed Wetlands from "anaerobic meltdown." Amazingly, wetlands don't stink. Despite a constant inflow of raw sewage, the oxygen pumping during daylight hours prevents anaerobic conditions and odors, especially at the surface, where it matters most. Root hairs also provide growth medium for beneficial microorganisms.

Treatment is quite effective. In one small wetland I made in Mexico (photo, facing page), the septic tank water going into it had 3,000,000 fecal coliform bacteria per 100 ml, and the wetland effluent just 50—the same as the tapwater in the house, and theoretically well within the standard for swimming. Nonetheless, disinfection may be required, depending on what you do with the effluent.

Ecological Machines are a design pioneered by John Todd.[s16] They are a wetland variation enhanced with the addition of a solar aquatic greenhouse, carefully crafted, balanced guilds of organisms, and attention to the disposition of each nutrient in the system. Ecological Machines range in flow from 500 to 5,000,000 gpd, require a land area sufficient for two day retention time, and cost as little as 30% less than a package treatment plant to build, 60% less to operate. These constructed ecosystems are especially appropriate in environmentally sensitive sites that don't offer much naturally occurring treatment capacity: for example, a ski resort surrounded by snow and rock, or cities. Some institutions choose this system for its teaching potential.

Note: Wetlands are proven for cleaning large flows of partially treated municipal wastewater and septic tank effluent. They are less common but successful for raw, unfiltered household greywater by itself; questions remain about clogging, high oxygen demand, and nutrient balance. Systems are reported to work with pantyhose pre-filtration or a large-area infiltrator at the inlet, but a septic tank is better proven. Design guidance from an engineer with Constructed Wetland experience is highly recommended.[s17,s18]

Automated Sand Filtration to Subsurface Emitters (DR)

Expensive and complex, but beloved by regulators. These systems start to make sense for new construction with at least 100–200 gpd (400–800 lpd) of greywater. Emitter cone installation does not get cheaper with larger irrigated area. Subsurface drip does, but more research is needed to establish whether subsurface drip works longterm with filtered but otherwise untreated greywater. ***This system and the next are the only acceptable ways to greywater turf.*** *With more than 500 gpd, Septic Tank to Subsurface Drip may be better (next section).*

Automated greywater to drip irrigation is the holy grail of greywater systems. Many have tried, few have succeeded, and none thus far have managed to open wide the gates between Americans' indoor water use and their lawns.

Reliable greywater to drip is elusive. Unless you enjoy destroying expensive equipment as a hobby, breathe deeply and buy a system from an outfit that knows what they're doing—if you can find one.[s3,s5,s7,s19]

Although the next big drought will no doubt spawn numerous new entries in this system category, ReWater Systems is the most solid representative in the US at the moment (Figure 8.9, Real World Example #5). They are the sole survivor of the numerous greywater system start-ups born from the big California drought of the 1990s. If you buy a system from an outfit that has been around for a long time, you'll hopefully get a more evolved design—and a better chance they will be there when the system needs an overhaul.

The ReWater computer-controlled, automatic backwashing, sand filter system handles every conceivable aspect of greywater irrigation automatically—it even coordinates freshwater irrigation. The idea is that it is your sole irrigation system.

One hundred twenty five gallons per day of greywater can water a 1,000–2,000 ft^2 lawn through emitter cones for roughly the same cost as sprinklers: about \$1.25/ft^2, installed (2007 cost). Sprinklers are cheaper on big turf areas up front, but cost more over the longterm due to higher water costs from overspray and evaporation losses.

Failure points for this type of system include inadvertent cutting of tubing when digging (easily done, easily fixed) and pump failure—about 4% fail early.

Subsurface drip tubing has clogged with sand-filtered greywater. There may be a fundamental incompatibility between the small orifices of subsurface drip emitters and greywater that has both high mineral and soap content, even after fine mechanical filtration. Or, these

FIGURE 8.9: AUTOMATED SAND FILTRATION TO SUBSURFACE EMITTERS SYSTEM

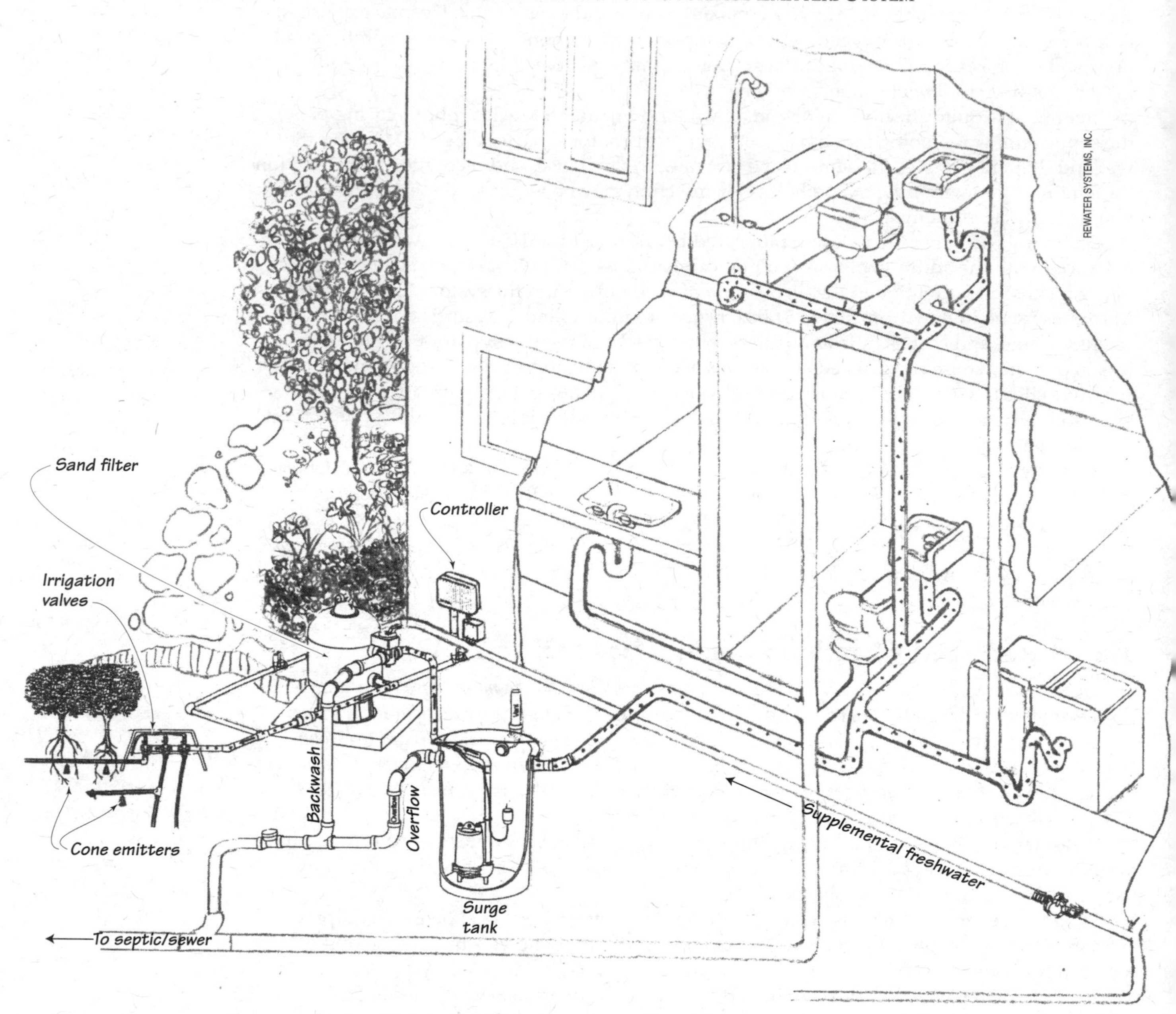

FIGURE 8.10: DISTRIBUTION CONE EMITTER

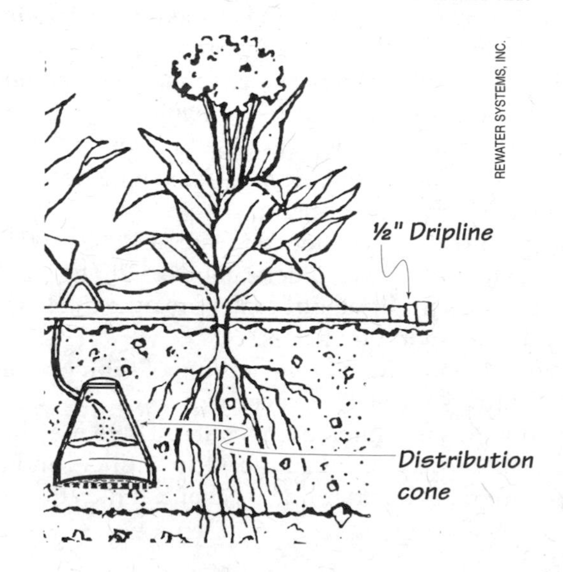

Installation of ReWater emitter cones for a new lawn. A tedious process, but at least they seem to work longterm.

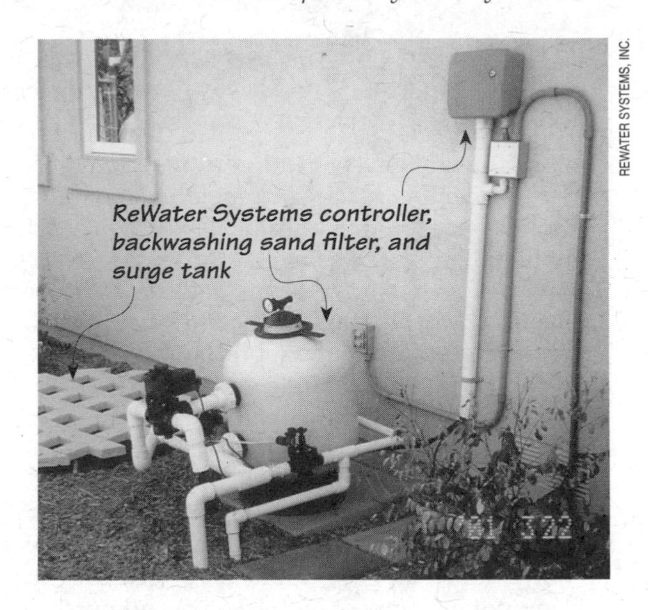

ReWater Systems controller, backwashing sand filter, and surge tank

failures may be due to absence of maintenance—more research is required. You can open up subsurface drip tubing by flushing it with acid, but this won't work if it is completely clogged.

ReWater has developed proprietary one-piece recycled plastic distribution cone emitters (Figure 8.10), approved by the Center for Irrigation Technology (as required by California greywater law). These cones have relatively large orifices that don't clog easily. They are about 3½" x 3½", and a 2½" x 2½" version is coming out.[m] They can be supplied with ¾", ½", or ¼" drip tubing. Roots grow into the emitters, but don't block the inlets thanks to air pruning (the reluctance of roots to grow more than a short distance straight up without soil, even though roots go hundreds of yards horizontally along irrigation pipes). According to ReWater, hundreds of thousands of these emitters are in the ground. The oldest, in use since 1990, are still working. Many of these systems are in San Diego, CA, which has some of the highest mineral content tapwater in the US.

Steve Bilson, the owner of ReWater, has been nothing if not tenacious, not only to simply survive so long in the industry, but in pursuing official recognition of greywater systems. With an effort of staggering proportions, he managed to get the US EPA State Revolving Fund (SRF) program to qualify ReWater systems for loans, which had previously only been available to cities for conventional centralized sewer systems, one of the century's greatest ecological disasters.[2]

His multi-year effort generated a 2" stack of paper and a rigorous life-cycle analysis showing that if you take everything into account, even as complex and expensive a system as the ReWater will pay for itself through savings on water, wastewater, and non-point source pollution prevention, despite lost interest, operations and maintenance, and system replacement costs.

According to Steve's analysis, sewer districts would save more water and money by installing gold-plated greywater systems in every new home than with the multi-million-dollar reclaimed water plants that the engineering lobbies are getting built instead. Interesting food for thought.

Caution: A halfhearted attempt at an automated greywater system is doomed. Components that work reliably with wastewater are expensive, but anything less will fail.

Septic Tank to Subsurface Drip (CDR)

Well proven for treatment, and proven, but less common, for reuse. Secondary treatment is overkill for a residential site unless it is required for some reason. This is the way to go for multifamily or institutional-scale flows. Expensive, uses a lot of electricity and materials, but is beloved by regulators. You can reuse blackwater with this system, obviating the need for dual plumbing. These are regulated under septic system regulations, not the greywater code, which is an advantage.

---

[m]***Metric:*** *9 cm x 9 cm, 6 cm x 6 cm.*

Here's a general recipe for success in reusing combined greywater and blackwater for irrigation:

- **Settling and anaerobic digestion**—Combined greywater and blackwater are filtered of most solids by settling in a strong, watertight septic tank (fiberglass or concrete). Anaerobic treatment reduces sludge, scum, and some nutrients (settling can be compromised in systems that bubble air through the septic tank itself).
- **Effluent filter**—A simple mesh filter protects the system from scum or sludge that might otherwise escape the septic tank (these are recommended for any septic system).
- **Secondary treatment (optional)**—Clarified septic effluent is stripped of biological and chemical oxygen demand (BOD and COD measure the amount of compost in the water) and nutrients in a well-designed, built, and maintained suspended growth aerobic treatment unit (ATU) or packed bed media filter.
- **Effluent pump in pump tank**—Pushes effluent through the filter and into the dispersal field.
- **Drip filter**—Mechanical filtration in a 100-micron, self-cleaning drip irrigation filter.
- **Subsurface drip distribution**—The effluent is distributed through well-designed subsurface drip irrigation.[s10] The whole drip system has to be a closed loop, the ends of which flush back to the pre-treatment or the pump tank to prevent clogging. This flush can be achieved through ball valves that are cracked open to leak continually and are fully opened at intervals, manually or by electrically operated solenoid valves.
- **Proper operation and regular maintenance**—More important than the specific choice of hardware is choosing hardware that can, and will, be serviced.

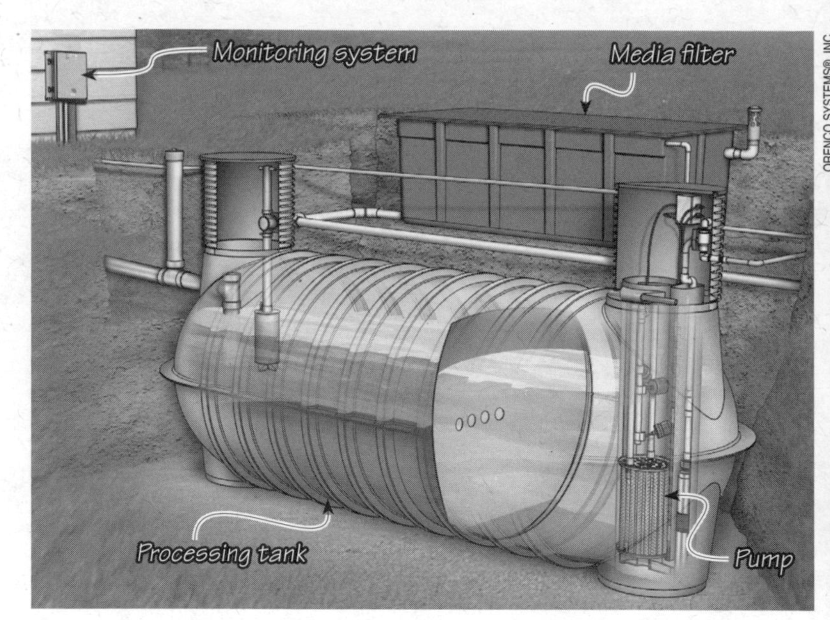

Orenco AdvanTex®—a popular secondary treatment system. *By using engineered textile instead of sand as the packed bed media filter, it takes about a quarter as much area in the yard.*

Drip dispersal of primary treated effluent is somewhat simpler and cheaper, but is allowed only in some states, may be problematic with clay soils, and requires more area.

Secondary treatment adds expense and complexity. The water is still an infection hazard, but it will not clog drip irrigation hardware as easily as primary treated (even finely filtered) greywater, especially in clay soils. It also allows higher loading rates and less separation to groundwater. Secondary treatment options are rapidly multiplying and evolving as large-scale solutions are adjusted to residential scale and vice versa.

In the ATU version of secondary treatment, air is bubbled through septic tank effluent to facilitate aerobic decomposition. This is the cheaper option.

In packed bed media filtration, effluent filters through textile, sand, or gravel (peat and foam have been used less successfully). Trillions of aerobic bacteria on the surface gobble the dissolved nutrients and anaerobic bacteria, yielding a sweet smelling, highly polished effluent.

Thus, for the high price of buying and maintaining these filters, pumps, and controllers, the dream of longterm, maintainable greywater to drip irrigation can be achieved. As toilet water needn't be excluded, you have more water available for reuse, don't need to pay for a backup wastewater system, and might be able to downsize or even eliminate the need for a freshwater irrigation system.

Orenco sand filter system being pressure tested *in Stinson Beach, CA. The pipes will be covered with sand and plants.*

As much as I dislike pumps, a pressure dosing system can drip-distribute water far more widely and evenly than a gravity system. Orenco Systems[s3] suggests using a high head effluent pump, protected by an effluent filter, to pressure dose subsurface drip irrigation lines. (For disposal, shallow Infiltration Galleys consisting of 12" PVC half-pipes in 10" *(25 cm)* deep trenches are an option.) People are also experimenting with ozone disinfection, which may someday permit this effluent to be used for sprinklers or garden ponds.

A residential Orenco Advantex packed bed media system can cost $12,000–$30,000 installed, plus about $200 a year for the required maintenance contract. An ATU unit could be a few thousand dollars less. Primary treatment only is not much less than that, because the cost savings on treatment are offset by more automation requirements and greater dispersal field size required.

Notes: Under some wastewater rules, it's better to call a wastewater reuse system a "dispersal" or "disposal" system and not an "irrigation" system. Despite movement toward acceptance of irrigation, a disposal mentality still hangs on. In most cases the system size and rate of application is based on soil acceptance formulas for the wettest month, not on irrigation need. This means that supplemental water will be required during the peak irrigation season.

This class of system is being pushed by industry profiteers and regulators as a universal substitute for septic systems, which would be a colossal waste of resources. Better to make simple, passive Green Septic systems (p. 67).

Marin's Stinson Beach, a long, very narrow, low sand spit with the Pacific on one side and an ecologically sensitive estuary on the other. The spit is covered with large houses, all on septic tank/sand mound systems, all working. As part of one of the most stringent monitoring programs in the country, Stinson Beach inspects all onsite wastewater treatment systems annually.

Cont. from p. 61

Chapter 9: Branched Drain Design

I recommend Branched Drains for most residential sized flows where the greywater sources are located above the irrigated area. This chapter and the next slice through the whole of greywater system design along the axis of this one system—an interesting exercise even if this particular system isn't for you. (This is another Art Ludwig original design.)

Most commercially sold greywater systems are elaborate affairs featuring filters, pumps, tanks, valves, and sometimes disinfection, electronics, etc. They cost $1,000–$30,000 (gasp!) for a single-family residence. Most are newly installed, abandoned, or failing to meet their original goals. However, the commercial systems account for only a tiny percentage of the greywater systems in use: 2% or less.

What about the other 98%? Many have worked for decades without the users even thinking about them. The vast majority are nothing more than a drainpipe pointing down the nearest hill. The classic Drain Out Back has some serious shortcomings, but its durability and spectacular simplicity give one serious pause for thought.

Lessons Learned from the Drain Out Back

From a holistic perspective, an overly complicated, expensive system is doomed from the start. At best, all it can do is shift the impact from waste of water to waste of resources used to produce pumps, valves, tanks, piping, and electricity.[15]

How about scrapping all this delicate, expensive technology, which is nearly impossible to make ecological, affordable, or durable, and instead concentrate on learning from and improving the humble Drain Out Back?

70 year old Drain Out Back in Southern California. *A shower drain waters an old-growth orange tree. Despite the fact that the outlet is quite close to a road that wraps around the mulch basin only a few paces away, no runoff is ever encountered. Leaves from surrounding oak trees automatically replenish the mulch.*

Advantages of the Drain Out Back

- **No filtration is necessary**
- **No pumping is necessary**
- **No surge tank is necessary**
- **Very little pipe is needed**
- **Little or no maintenance is required**
- **Economic and ecological costs are low**
- **Anyone can build it**
- **The failure rate is low** (that is, they are little worse in year 100 than year one)
- **They last forever**

Drawbacks of the Drain Out Back

- **Low reuse efficiency**—With all the water dumping in one spot, the result tends to be one overwatered patch and everything else underwatered. Usually nothing is planted to utilize the water, so the reuse efficiency is zero. Even when there are plants at the outlet, they might utilize only a fraction of the water available. In a few lucky instances there is a conveniently situated fruit tree, which grows until its water need equals the flow, in which case the reuse efficiency is high.
- **Poor sanitation**—Without a basin to contain it and mulch to cover and slow it, greywater applied to the surface could flow into a creek or onto the street, especially during rain. In older installations near natural waters, the pipe typically discharges directly into them (see Error: Greywater Staying on the Surface, Chapter 11). Without mulch or soil covering it, greywater could be lapped up by dogs, played with by children, or prowled for food bits by vermin. (This litany sounds worse than it really is in most industrialized world installations. A thin sheet of wastewater just flowing over the ground for even 20 paces receives spectacularly high treatment by the same beneficial bacteria that live in soil.[3] But in non-industrialized world conditions, especially in shanty towns where people live in appalling intimacy with their greywater, it is just as bad as it sounds above.)

- **Soil overload/poor aesthetics**—With a typical Drain Out Back, a mucky, grey-white material on the ground indicates where the soil's purification capacity is overloaded. This patch can measure from a few inches to several feet in length. If the ground slopes so the greywater runs off and infiltrates the surrounding soil, there is usually no odor unless you stick your face in it. If the greywater pools at the outlet, odor is likely and mosquitoes are possible. Some plants may be harmed by root suffocation.
- **Illegal**—Though common throughout the world, the Drain Out Back is generally illegal because the effluent is exposed on the surface (actually getting busted, however, is virtually unheard of). Drain to Mulch Basin, a more refined version, is permitted for new construction or remodeling in a small but growing handful of jurisdictions including Arizona, New Mexico, and Texas.[4] Greywater Furrows are like Drains Out Back with greater irrigation efficiency and less anaerobic muck. (Drain to Mulch Basin and Greywater Furrows are superior variations that can be substituted virtually anywhere a Drain Out Back could go.)

Branched Drain ancestor. *A cement channel distributes greywater on a steep, seasonally dry slope in Cuarnevaca, Mexico. This system distributes greywater more widely than if it was just dumped on the ground. However, due to imprecisely formed splits, numerous cracks, and blocking from roots and leaves, it is not as predictable as a Branched Drain network. Shaping the splits precisely with a form would help.*

Branched Drains to the Rescue

A Branched Drain system solves most of the drawbacks above while retaining most of the advantages. Like the Drain Out Back, it has no pump, filter, surge tank, electronic controls, or any opening smaller than 1", and has maintenance intervals of a year or more.

Unlike the Drain Out Back, it can automatically disperse greywater to several trees with satisfactory efficiency, it is much more sanitary, and it is possible that you could get a permit for it. The difference is that it *splits* the greywater in a branching network of pipes, then *contains and covers* the greywater in the landscape by means of mulch-filled basins.

Split the Flow

The most intractable problems of the Drain Out Back stem from an unmanageably high flow to one place. The Branched Drain system addresses this by splitting the flow (Figure 9.1).

FIGURE 9.1: FLOWS SPLIT VIA BRANCHED DRAIN (PLAN VIEW)

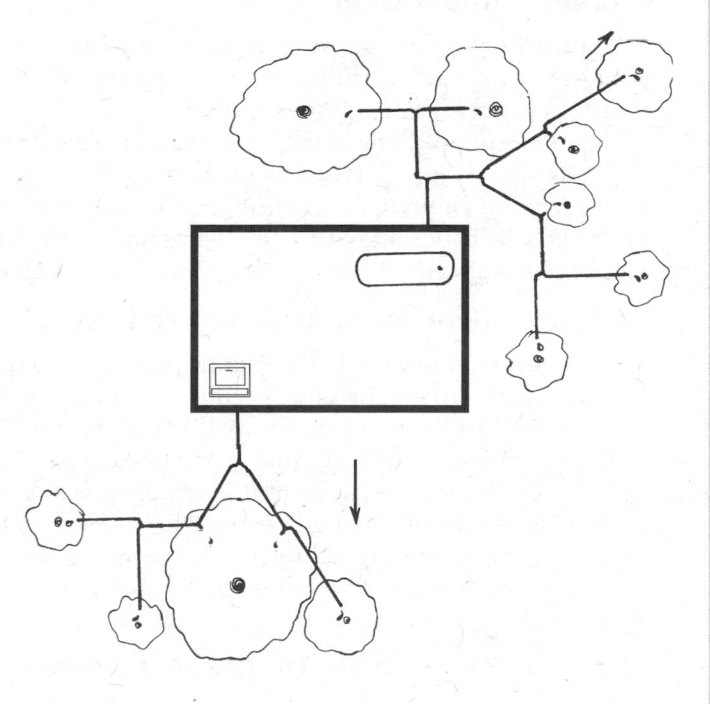

Contain and Cover the Flow

Mulch basins contain and cover the flow (Figures 9.2, 9.3). They are common in horticulture and could hardly be simpler to make and maintain. Don't let this fool you. Though nature takes care of their inner workings, these are fantastically complex biologically, far more than a municipal sewage treatment plant. What's more, the treatment level that mulch basins provide is much higher.[3] Finally, instead of consuming copious electricity and chemicals to pollute natural waters and create piles of toxic sludge,[2] mulch basins run on sunlight and yield drinkable groundwater and fresh fruit. Are you convinced yet? (Mulch Basin Design is covered in depth in Chapter 5.)

Greywater outlets into the basins can be arranged so the water falls through the air for a few inches before disappearing under the mulch (simplest, lowest maintenance, most durable). Or, they can be fully enclosed in chambers under mulch (more legal, more sanitary) or in chambers under soil (most legal, most expensive). These options are fully described later under Branched Drain Outlet Design.

Continue p. 100

FIGURE 9.2: GREYWATER CONTAINED AND COVERED IN A BRANCHED DRAIN-FED MULCH BASIN (ELEVATION VIEW)

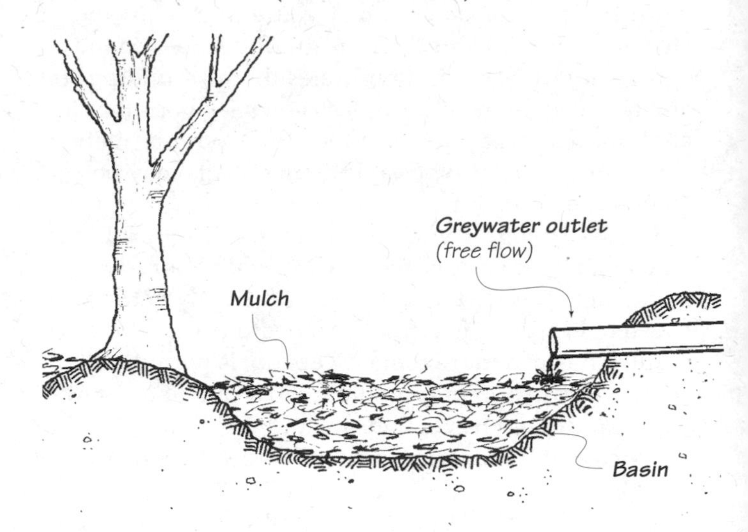

Advantages and Disadvantages of Branched Drain Systems

Overall I'd characterize them as "surprisingly involved to design and install optimally, with surprisingly little to do afterward." The upfront investment (primarily labor) is substantial but then it's mostly over, possibly for the life of the house. In contrast, most other systems cost more, lack the opportunity to save much by doing the labor yourself, and require significant ongoing inputs of electricity, maintenance, and system replacement—or, they require much more effort in use.

All versions of the Branched Drain system are permitable in Arizona, New Mexico, and Texas. If you use one of the subsurface versions, it is possible to get a permit under the CPC/UPC. If you're dealing with permits, I suggest you get our *Builder's Greywater Guide.*[6]

FIGURE 9.3: FLOWS SPLIT BY BRANCHED DRAIN NETWORKS (PLAN VIEW)

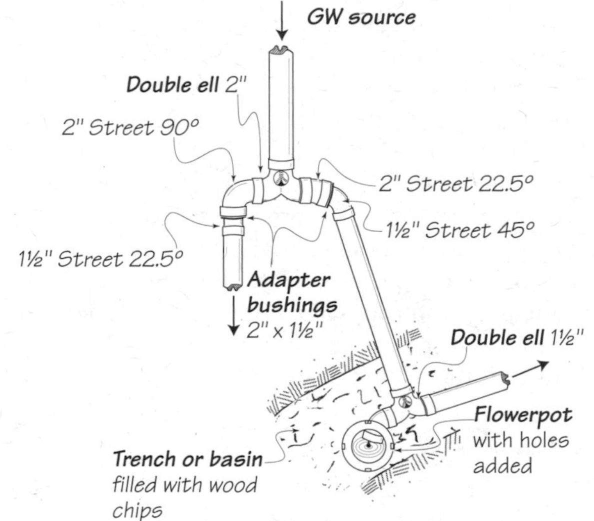

Limitations of the Branched Drain System

Carefully consider the limitations below and make sure what you are setting out to do is possible.

Cannot Deliver Water Uphill

If the irrigation or treatment area is located above your greywater sources, you're stuck. Look at the System Selection Chart for other system options.

Non-exception: Conceivably, a washing machine or an effluent pump could deliver greywater to an uphill, gravity flow Branched Drain network. But in both these cases you'd most likely be better off with a smaller diameter, more easily laid network directly pressurized by the washer or effluent pump.

The Ground on Your Site Needs to Slope at Least ¼" per Foot (2%)

The steeper your lot and the higher your house plumbing is above ground level, the less painstaking it is to lay the pipe. If your ground slopes 1" per foot *(8%)* or more, you can pretty much just dig a trench, throw the pipe in, and throw the dirt in after it. (However, with very long pipe runs the water may run ahead of solids, leading to clogging.) If it slopes ¼–½" per foot *(2–4%)*, it is a tedious multi-step process to level the pipe. If your lot slopes less than ¼" per foot—wow, it's going to be rough. In this case, the farther from the house you go, the deeper your lines end up, and you can't irrigate very far from the house. You have to double the design effort and redouble the installation precision. If you have a lot of water to get rid of and low perk soil...it may not be feasible at all.

The Pipes Must Be Sloped to Exact Tolerances

The tricky part of Branched Drain systems is designing and installing them so the pipes slope continuously and correctly downhill, within close tolerances of ¼" or less *(6 mm)*.

TABLE 9.1: ADVANTAGES AND DISADVANTAGES OF BRANCHED DRAIN GREYWATER SYSTEMS

Parameter	Most commercial systems	Drain Out Back	Branched Drain with free flow outlets	Branched Drain to subsurface chambers
Filtration	Needed	Not needed	Not needed	Not needed
Pumping	Needed	Not needed	Not needed	Not needed
Surge tank	Needed	Not needed	Not needed	Not needed
Amount of plastic	High	Low/zero	High	Higher
Maintenance	High	Low/zero	Very low	Low
Economic cost	High	Very low/ zero	Low	Pretty low
Ecological cost	High	Low	Low	Pretty low
Ease of building	Moderately difficult	Very easy	Requires careful pipe leveling and design	Requires very careful pipe leveling and design
Failure rate	High	Almost never	Extremely good so far	Very good so far; chambers tend to fill with debris, adding maintenance, or in the absence of maintenance, a possible failure
Life expectancy	Low	Forever	Probably decades	Probably decades, but with more maintenance
Reuse efficiency	5–90%	Can be 80% but usually zero!	50–80%	50–80%
Sanitation	Variable	Poor	Excellent	Impeccable
Soil loading	Variable	Poor	Good	Fair
Aesthetics	Variable	Poor	Good	Excellent; all is hidden
Legal?	Variable	Never	In AZ, NM, and TX	Yes
Flexible?	Varies	No	Not very	Not very; chamber must be moved as well as pipe

Hiring a plumber to get the inside pipes right is a good idea, but plumbers aren't that great in the yard. First, they are expensive. Second, they are easily disoriented in the garden. Third, plumbers tend to be conservative and tradition bound, and you might have a hard time coaxing them to adhere to your design. A landscaper or a jack-of-all trades might be a better bet, but you are likely to have to do much of the work yourself, or at least supervise it. Unless you are naturally detail oriented, you may have a hard time warming to the sometimes painstaking task of getting those darn floppy pipes to slope just right in the shifting and settling soil.

Exception: If you have lots of well-situated fall, getting slopes right is a breeze.

The Pipes Must Slope Downhill Continuously

The pipes cannot drop down and then back up again: for example, plunge to exit under the foundation and then pop up to the surface, or dip to cross a gully.

Possible exception: If there is a large, unvented elevation drop (10–20', *3–6 m*) on the line, the greywater pressure should blast through any clog in a U-shaped portion. But, water pooling in the U goes septic after a while, and when fresh greywater pushes it through, it comes out stinking. Set a system up this way only if there *truly is no alternative*. Don't even think of it if you don't have a lot of pressure—it is sure to clog.

Impractical for Lawns or Small Plants

Beside the other reasons not to greywater lawns (see A Note on Lawns, Chapter 2), the Branched Drain system, with only a few dozen outlets maximum, is poorly suited to irrigating thousands of tiny individual plants. It works best with trees.

Exception: If your only goal is to get rid of the water, it may be appropriate to use plants whose claim to fame is sucking up water, regardless of their size.

Poorly Suited to High Flows

There may be an elegant way to reuse more than a few hundred gallons a day through a Branched Drain network, but I haven't thought of it. The amount of pipe and the area required to deal with a single large source soon seems overwhelming. This limitation is shared by all but a few greywater system types. The average, mildly conservative, industrialized world household generates 35 gal/person/day *(130 L/person/day)*, which is well within the comfortable range for the Branched Drain system (there is no lower limit).

If you are faced with a huge water flow and a similar irrigation need, the economics of other, more elaborate systems become more attractive.

Difficult to Alter

With this system, there is no easy way to supply a small amount of water to a young tree and then several times more water once that tree is large, or to shift from watering deciduous trees to evergreens when the deciduous trees drop their leaves. You make your best approximation when you design the system, and that's pretty much it.

Exceptions: It is easy to add to the ends of lines, or shorten outlet extensions. For example, you can design a system to accommodate tree growth by putting the flow splitter some distance from the trunk. An outlet extension from the flow splitter to a young tree can be removed when the roots have grown out farther, or an additional split can be added to send the water to two places in the (now) extensive root zone.

If one side of a flow splitter is the final outlet (as in the photo on p. 85), you can plug it and the water will continue on unsplit. This could be used to bypass a deciduous fruit tree that has lost its leaves and send the water on to a thirsty evergreen. When the deciduous tree leafs out, remove the plug and clear the opening.

Unproven for More than 16 Outlets from One Source

It is certainly possible that more outlets could be served, but it hasn't been tried to my knowledge. With ordinary flows, 16 is a reasonable limit. With pulsing output from a dipper box, 64 outlets would likely work (see Branched Drain Variations, Improvements, and Unknowns, end of Chapter 10).

Branched Drain System Design

After you've figured out what plants you want to irrigate with a Branched Drain distribution system, you're ready for the next steps: determining how to split the flow, designing the branching geometry, choosing the outlet types, and designing the mulch basins.

Ways to Split the Flow

There are several ways to split the flow (Figure 9.4). Let's consider them in turn:

1. **With a double ell or other tee pipe fitting that splits the flow in two**—By then splitting it again, and again, and again, you can end up with as many as 16 outlet streams from one flow, in a branching network. This is the best way to split the flow in most cases.
2. **With a dipper distribution box**—These normally split the flow in four in one pass, but could possibly be used to split the flow into as many as eight outlets.
3. **By manually moving a flexible outlet extension from place to place**—This is excellent to use after the collection plumbing is done, while you're thinking about a permanent distribution system. (See Movable Drain, Chapter 7. Flexible PVC is also useful for transporting rainwater from downspouts to your greywater system to flush salts from the soil.)
4. **By not combining the flows in the first place**—Each fixture or set waters its own area.
5. **By any combination of 1, 2, 3, and 4.**

The first design decision is whether to go with one combined line, or separate the flows first by not combining them. This decision determines the layout of the collection plumbing in the house and is not easy to change. The pros and cons of each approach are described in Lump or Split the Greywater Flow, Chapter 3. *(Note: If you want Multiple Greywater Zones—the topic that follows Lump or Split in Chapter 3—you'll probably want to lump the flow.)*

FIGURE 9.4: WAYS TO SPLIT THE FLOW
(PLAN VIEWS)

Darker shaded areas have greater irrigation need. Unshaded areas are irrigated occasionally or not at all. (Septic/sewer lines not shown for clarity.)

2" GW distribution pipe
1½" GW distribution pipe
3-way diverter valve
Greywater outlet
2" GW collection pipe
1½" GW collection pipe

A: FLOWS "SPLIT" BY NOT COMBINING THEM IN THE FIRST PLACE, THEN BY FLOW SPLITTERS

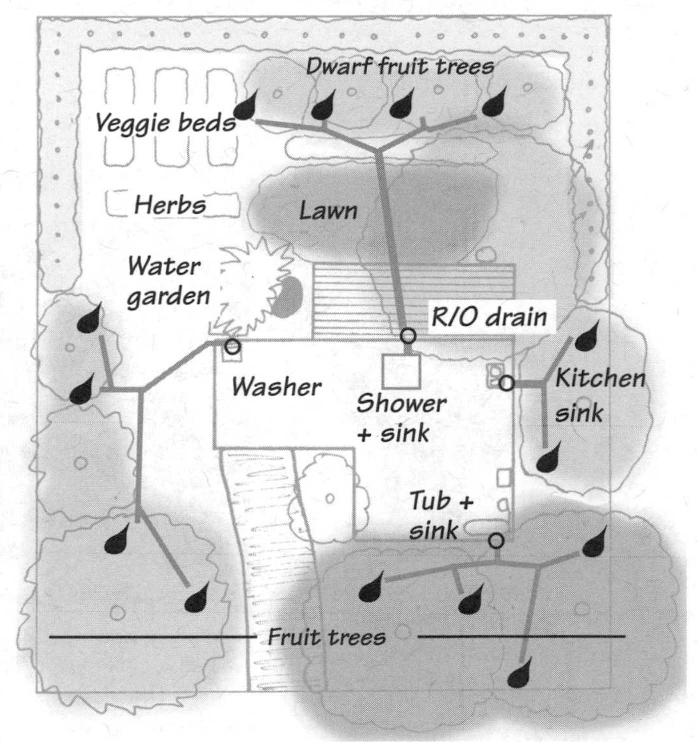

B: COMBINED FLOW SPLIT INITIALLY BY A DIPPER BOX, THEN BY FLOW SPLITTERS

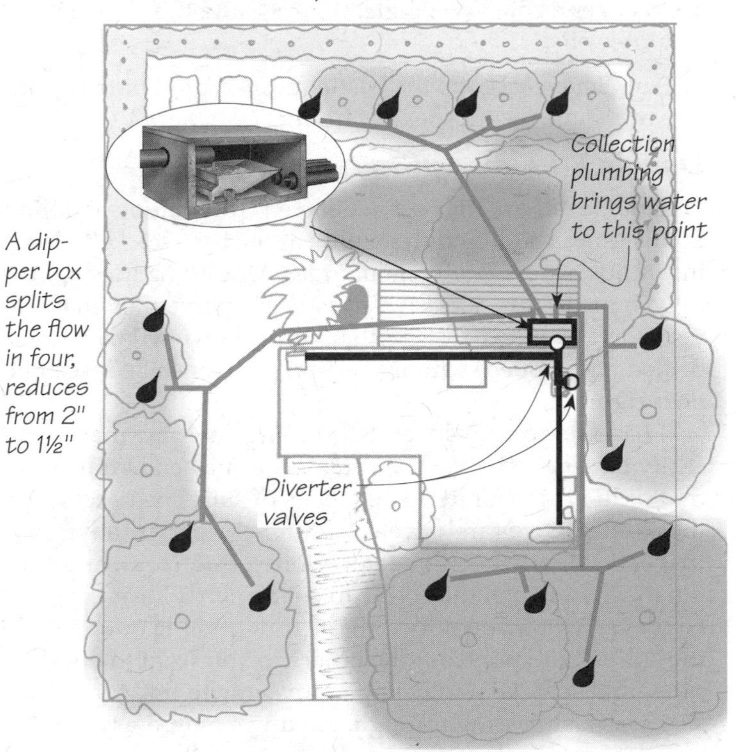

C: COMBINED FLOWS, SPLIT WITH FLOW SPLITTERS ONLY

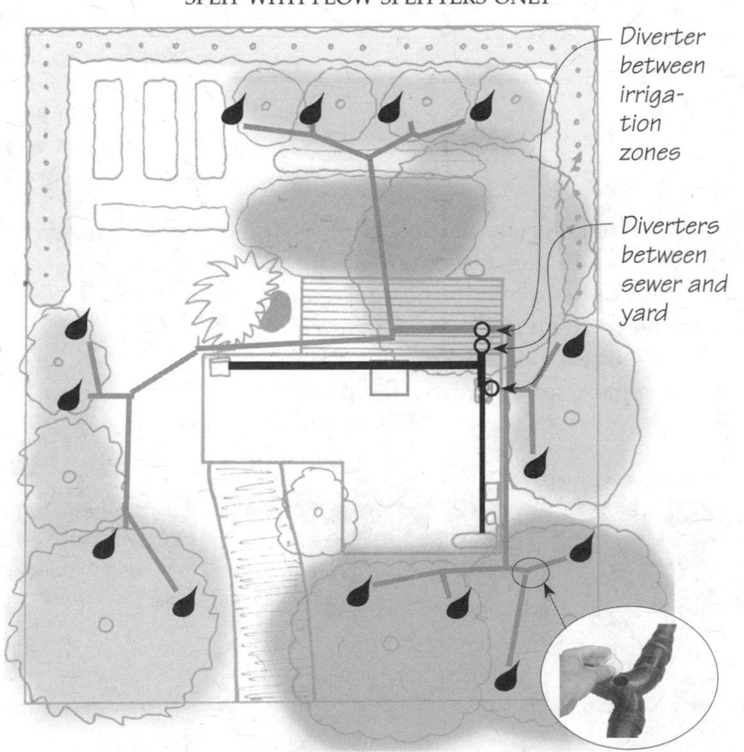

D: UNCOMBINED FLOWS "SPLIT" BY MANUALLY MOVING AN OUTLET EXTENSION[s6]

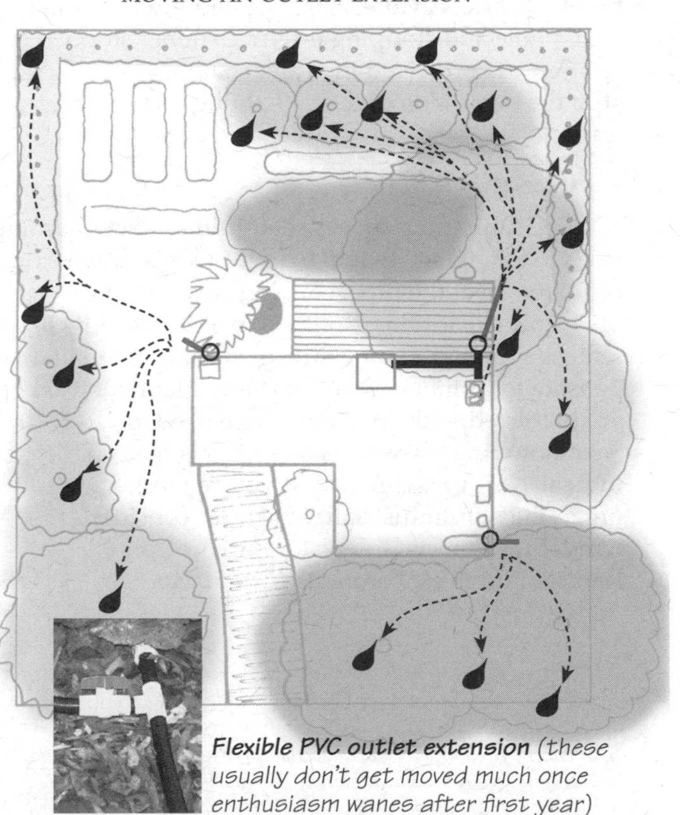

Flexible PVC outlet extension (these usually don't get moved much once enthusiasm wanes after first year)

Parts for Splitting the Flow

Flow splitters and wastewater bends are not available in sizes smaller than 1½", so that is the lower size limit, at least until the last split. Some plastic is saved by reducing to 1" pipe for the final outlet extension after the flow splitter (photo on p. 84 shows a reduction to 1"; see also Smaller Pipe or Other Types of Pipes, in Chapter 10).

A system uses more pipe and more fittings than you'd expect. Branched Drain networks also require large quantities of rarely used fittings such as:

- street bends of every angle: 1/16, 1/8, 1/6, and 1/4 bends (22.5°, 45°, 60°, and 90°, respectively)
- double ells

You'll have to go to the best-equipped plumbing supply house around, and even they may not have the most exotic fittings in Table 9.2.

Double Ell Flow Splitters

Double ells are the best available flow splitting coupling at this time. You'll find them only at the best-stocked plumbing supply houses, or by mail order. Modifying flow splitters to gain inspection access will vastly simplify maintenance. If you don't want to hassle with it yourself, we sell them already drilled and tapped, with inspection plugs installed (see photo near right).[s9]

I have a strong aversion to building anything that cannot easily be serviced. I've never made a Branched Drain network that didn't incorporate some means of simplified access to the splitters. I've rarely seen a splitter clogged by anything that came down the pipe*—more often dirt, rocks, and leaves fall in when I open the access to confirm that it isn't clogged! However, when you notice all the flow coming out one side of the splitter and none out the other, do you want to dig up the pipes and saw them in half, perhaps only to find that a banana slug was just passing through, making a temporary dam?

Although it is possible to make an installation with no way to inspect the inside of the flow splitter, I don't think it's a great idea. Installing your own threaded inspection plugs is a bit of a hassle. An alternative possibility is drilling a plain 1¼" *(3 cm)* hole and stopping it with a rubber stopper. If your flow splitters are protected in outlet chambers, you don't need to stop the inspection hole at all, though a bit of water might slop out.

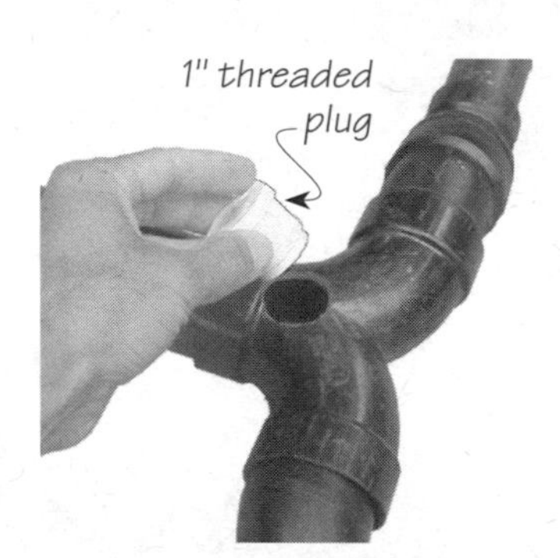

A double ell retrofitted with inspection access. *The access is sealed with a threaded plug.*[s9]

Cutaway view of a double ell, *a flow-splitting fitting that is largely self-cleaning with no filtration. This one is part of the first known Branched Drain network. I valiantly tried to clog it with pure kitchen sink water—took the sink strainer out and pushed bowls of soggy granola and pot scrapings down the pipe for half a year. The p-trap clogged every other day, but the Branched Drain network never did.*

PVC Sanitario

PVC sanitario is an eggshell-thin system of drainpipe with a small variety of loose fittings to join them. To make up for lack of fittings other than 90° and 45°, resourceful Mexican plumbers heat the pipe, creating clean bends, as well as bell ends in lieu of couplings. Instead of double ells for flow splitters, we successfully used sanitary tees, which in this system are almost as straight as a vent tee. These are definitely not optimal but they're working. To compensate for thinness and consequent tendency to break (under a path, for example), the pipes are protected with concrete where needed.

See photos of systems made with this material in Real World Example #3. If you're working in the non-industrialized world, consider Landscape Direct, Greywater Furrows, and Radical Plumbing before PVC sanitario.

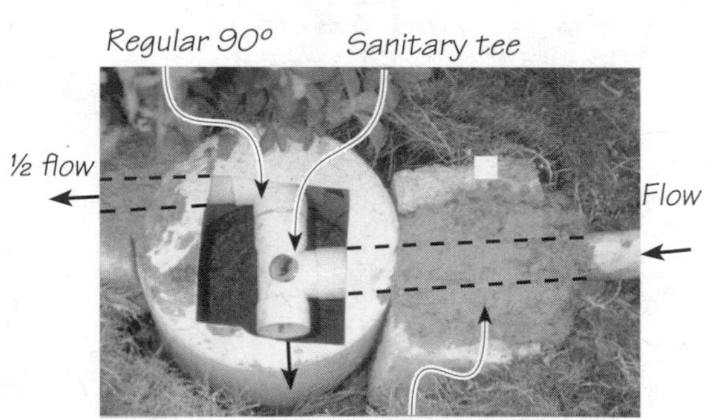

A Branched Drain of PVC sanitario. *The flow splits inside an outlet chamber; half dumps out and the rest flows on.*

**Double ells from at least one manufacturer have ragged injection molding ridges right on the smooth face where the flow is supposed to split. If you get stuck with these, grind down the ridges with a rotary file.*

TABLE 9.2: PARTS FOR BRANCHED DRAIN DISTRIBUTION PLUMBING

Component/Photos	Application/*Material*	Typical cost+	Quantity used	Sizes used	Normally used for/*available from* (DWV = drain-waste-vent)
Diverter valves			0–3		
3-way valve	Divert water between zones in yard/*PVC*	$50	0–3	1½" + 2" 2" + 2½"	Pools and spas/*pool suppliers*[s2] *and oasisdesign.net*
Pair of RV dump valves	Alternative to 3-way valve/*PVC*	$25	0–6	1½–2"	Recreational vehicles/*RV shops*
Backflow prevention valves					
Backwater valve	A CPC/UPC legal requirement to keep blackwater from backing up and going out into greywater systems*/*PVC, brass*	$25	0–1	2"	Backwater/*well-stocked plumbing supply house*
Swing check valve	A cheaper substitute for a backwater valve/*PVC*	$25		2"	Swing check valve is used for potable water/*well-stocked plumbing supply house*
Dosing devices			0–1		
Dipper box	The dipper box can serve as dosing device, flow splitter, pipe size reducer, and greywater meter[28] (see Dipper Box, p. 82)	$130 + trans-port		2–4" in, 1½–2" out	Dosing septic tank leachfields. Must be customized to accept small pipe/*call manufacturer*[s20]
See Figure 4.5 Dosing siphon	Dosing device			2"	Dosing septic tank leachfields. Must be customized to accept small pipe/*call manufacturer*[s3,s4]
Flow splitters			1–15	1½", 2"	
Double ell	Best way to split flow in two/*ABS*	$3.50			DWV/*order pre-fitted with cleanout plug from oasisdesign.net*[s9] *or make them from parts from a well-stocked plumbing supply house*
Vent tee	Best substitute if double ell is not available/*ABS*	$9			DWV/*well-stocked plumbing supply house*
PVC sanitario tee	Workable flow splitter available in Mexico/*PVC*				DWV/*any hardware store in Mexico*
Sanitary tee	Possible but not recommended flow splitter/*ABS*	$3.75			DWV/*most hardware stores*
1" threaded plug	Plugs access to flow splitters/*PVC*	$2.75			DWV/*most hardware stores*
Reducers			0–2		
Reducing bushing	Reduce from 2" to 1½" pipe/*ABS*	$2	2	1½" × 2"	DWV/*most hardware stores*
Reducing bushing	Adapt from ABS to 1" flexible PVC/*PVC*			1" × 1½"	
Reducing coupling	Reduce from 2" to 1½" pipe/*ABS*	$2			DWV/*well-stocked plumbing supply house*
Bends			10–20	1½", 2"	
¼ bend 90° street elbow	Turn at flow splitter/*ABS*	$3			DWV/*most hardware stores*
¼ bend 90° extra long elbow	Horizontal 90° turn in mid-pipe/*ABS*	$4.25			DWV/*well-stocked plumbing supply house*
1/16 bend 22½° street elbow	Slight turn/*ABS*	$3			DWV/*well-stocked plumbing supply house*
¼ bend 90° elbow	Turn at flow splitter/*ABS*	$2.75			DWV/*most hardware stores*
¼ bend 90° extra long street elbow	Deluxe way to make 90° turn out of flow splitter in installation with very little fall/*ABS*	$2.25			DWV/*well-stocked plumbing supply house*
⅙ bend street elbow	Turn at flow splitter/*ABS*	$7.25			DWV/*well-stocked plumbing supply house*
⅛ bend 45° elbow	Turn in mid-pipe/*ABS*	$2.75			DWV/*most hardware stores*
⅛ bend 45° street elbow	Turn at flow splitter/*ABS*	$3			DWV/*most hardware stores*
Cleanouts, etc.			1–4	1½", 2"	
Cleanout wye w/plug	Cleanout/*ABS*	$2.75			DWV/*well-stocked plumbing supply house*
Fitting cleanout w/plug	Cleanout/*ABS*	$1			DWV/*most hardware stores*
Long turn tee wye	Cleanout/*ABS*	$4			DWV/*most hardware stores*
Pipe, etc.				1", 1½", 2"	
Pipe	Piping/*ABS*	$1			DWV/*most hardware stores*
Coupling	Joining straight pipes/*ABS*	$1.25			DWV/*most hardware stores*
Repair slip coupling	Splicing into already-glued system/*ABS*				DWV/*well-stocked plumbing supply house*
No-hub connector	To join pipe, street fittings without glue so they can be removed/*Stainless steel, rubber*	$4			DWV/*well-stocked plumbing supply house*
Spa-flex	Movable outlets/temporary systems/*PVC*	$2.25			Hot tubs/*pool and spa supply house*

+*Costs are for 1.5" pipe and fittings. Fittings for 2" pipe typically cost 50% more.*

**But not required to keep blackwater from backing up out your shower drain! Likely to clog, so install it with no-hub connectors and/or a cleanout.*

Dipper Box or Dosing Siphon (Optional)

Most greywater systems use surge tanks to flatten peaks out of the flow. But Branched Drains do well with flow spikes—so much so that a device is sometimes added to transform a low flow into strong, intermittent pulses. Do you need such a thing? You do if:

- **A large percentage of your greywater dribbles out at a very low flow rate**—The dipper is designed for the slow drool that accounts for much of the flow out of a septic tank.
- **Daily flow is high**—The dipper splits the flow in four (or more) in one step, while adding such a peak to the flow that, even cut in quarters, the rate is higher than what you started with. This is one way you could cut a flow into more than 16 streams (see More than Four Levels of Splitting, Chapter 10).

My favorite "surge creator" is a custom version of the Polylok dipper.[s20] It comes standard with flexible plastic knockouts well-suited for 3–4" leachfield pipes held in place by earth in the septic tank installations it was designed for. Unfortunately, these connections don't hold small pipe steady and are especially awkward in the aboveground installations preferred for greywater. Dipper boxes are made one at a time. If your local supplier is cooperative, you can have your dipper made with a 2" inlet and overflow, and four 1½" outlets, in the form of PVC or ABS couplings. This makes a much cleaner installation. You can use one or all of the outlets. An overflow line is a good idea if the dipper is someplace where it would make a mess if greywater overflowed out the top; if it is outside, an overflow line isn't necessary. Theoretically you could have a dipper made with just about any number of outlets, including outlets of different sizes to obtain controlled splits into different-size flows. Figure 9.4b shows a sample layout using a dipper box to split the flow into four in one step. You can also get a standard dipper and adapt the pipes yourself with cement.

If you have to measure your greywater flow for some reason, you can use a dipper to do so.[28] The dipper costs about $130 (plus transportation) and uses 8" *(20 cm)* of fall. It is a few feet on a side and weighs 360 lbs *(160 kg)*—UFF!

Dosing siphons[s4] are a related technology. They are not generally designed to split the flow, only to collect a lot of water and then distribute it all at once. The water can easily sit in them long enough to become anaerobic, and unlike the dipper, dosing siphons collect crud, which must be cleaned out periodically. They also cost more fall—the height difference between inlet and outlet is typically a few feet. Dosing siphons are described in Surge Capacity, Chapter 4.

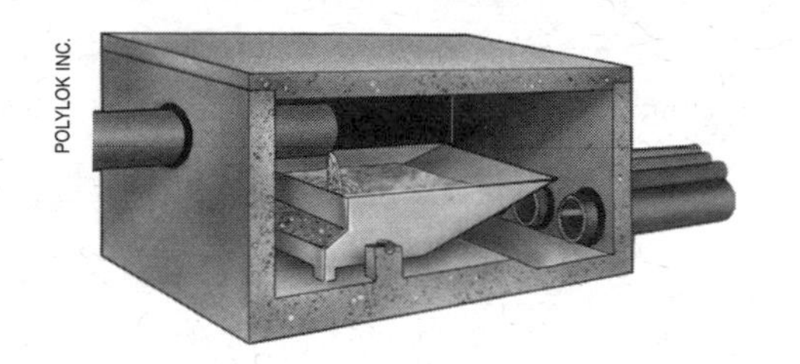

A Polylok dipper box.
When the tray fills, it dumps all the water and solids, and then pops back up, as shown in the sequence below. Have your outlet couplings installed so the bottom quarter of the pipe is below the floor of the dipper (as at left). This will provide a stronger flow to downstream flow splitters at the end of the cycle.

Figure 9.5: Dipper Distribution Box Dipping Cycle

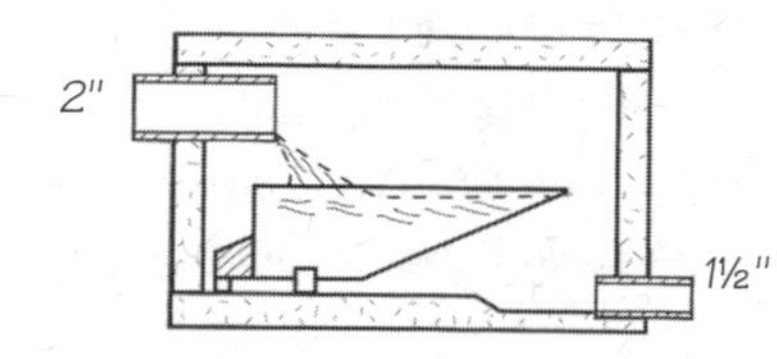

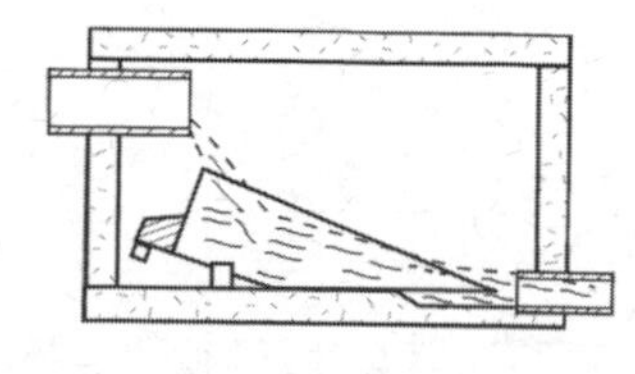

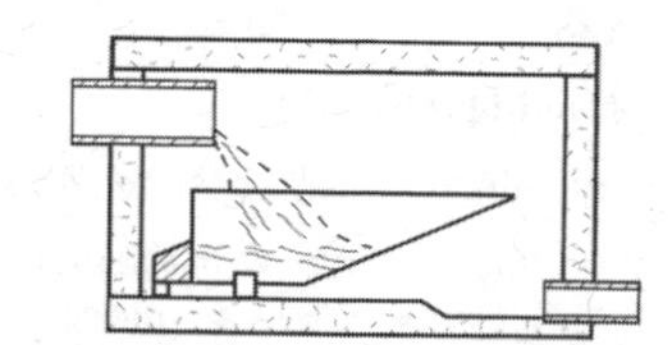

Branching Geometry Options

Even with the greywater coming from a known point and going to other known points, there are still a few different options for setting up the geometry of the branching network:

FIGURE 9.6: BRANCHING GEOMETRY OPTIONS

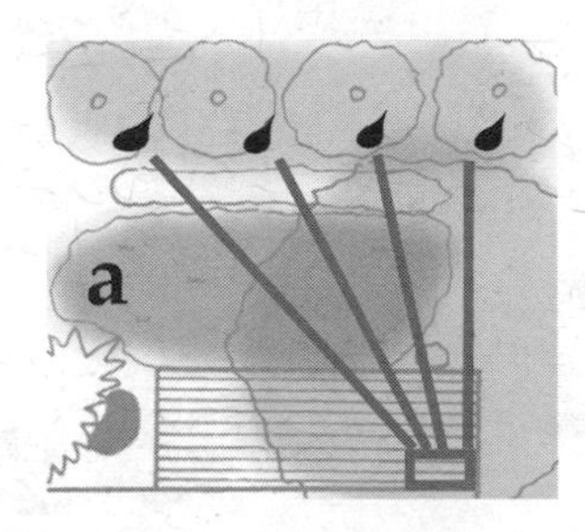

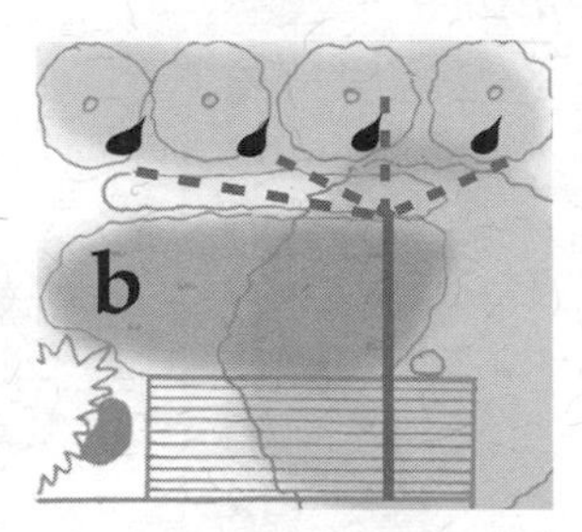

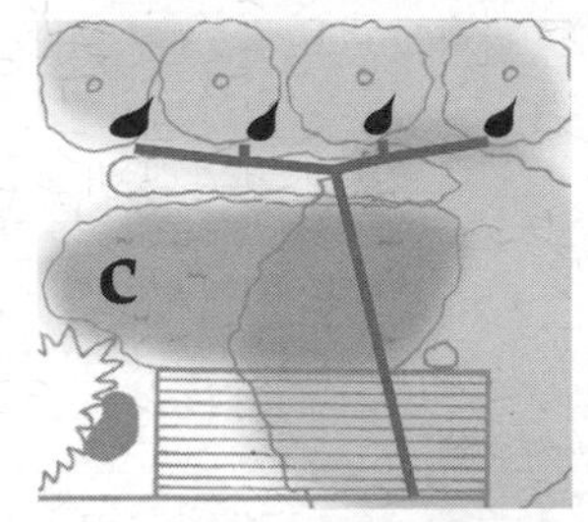

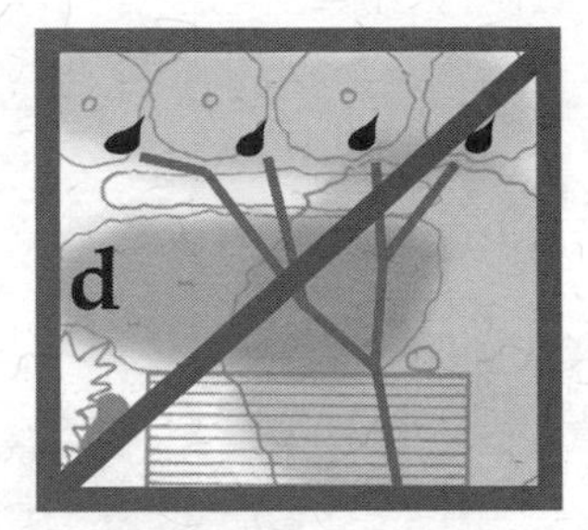

The same outlet points receiving the same amount of water from the same source point using four different branching options: A) a dipper box makes the split; B) a manually moved flexible outlet extension serves four points (if someone remembers to move it); C) branches with one short leg, and all turns at the flow splitters; and D) branches of somewhat equal length with turns mid-pipe (the only one of these options **not** recommended).

All the options in Figure 9.6 are good except d. Do c instead, with all the turns at the flow splitters, and straight runs of pipe otherwise. I'm attracted to this approach for a few reasons:

- **It takes fewer glue joints**—By connecting the bends at the flow splitters, you can use street bends (instead of a hub at each end, they have one hub end and one pipe-size end; see Figure 9.7). This cuts the number of glue joints in half. Fewer glue joints means less use of noxious glue and a smoother water path.*
- **It's easier to find and fix clogs**—Clogging is most likely in bends and connections. If the bend is right by the flow splitter, you can find and most likely unclog it via the access plug in the flow splitter. (The figure Sample Plot Plan for Permit in our *Builder's Greywater Guide* shows how not to do it: a plan for a Branched Drain network with numerous 90° bends in mid-pipe. Fortunately, it wasn't actually built this way.)
- **It's easier to find piping**—If the pipe runs are straight from flow splitter to outlet, locate the flow splitters (on a site map, and/or by digging), and you know where the pipe runs are.
- **It takes less plastic**—This layout usually yields the most pipe- and coupling-efficient, least expensive system.

Unless the flow is very small, there will be several splits. First, work these out on paper. Next, check out your plan on the ground by marking where you want the water to come out of the house and where you think the outlets should be. Then, lay out the pipes on the surface and connect them with fittings—don't glue or cut anything yet!

Now, repeat the paper and physical layout steps above, trying different approaches and refinements until you feel you've optimized everything. Don't try to do *all* the fitting-by-fitting detail or to locate the pipes to the *inch*. Get a representative fraction of the fittings worked out and locate the pipes and outlets to within a foot or two. During installation (Chapter 10), as you lay the pipe from top to bottom, the pipe and fittings themselves lead you to the resolution of that last bit of detail.

FIGURE 9.7: "STREET" VS NORMAL FITTINGS

¼ bend 90° elbow

Hub

¼ bend 90° street elbow

Hubless, pipe-size end

Street fittings have one pipe-size end and one coupling-size end.

On the down side, street fittings apart from 45°s and 90°s are harder to find and more expensive than their symmetrical counterparts.

Reductions

Reductions in Branched Drain distribution systems require special treatment in order to not become clogs:

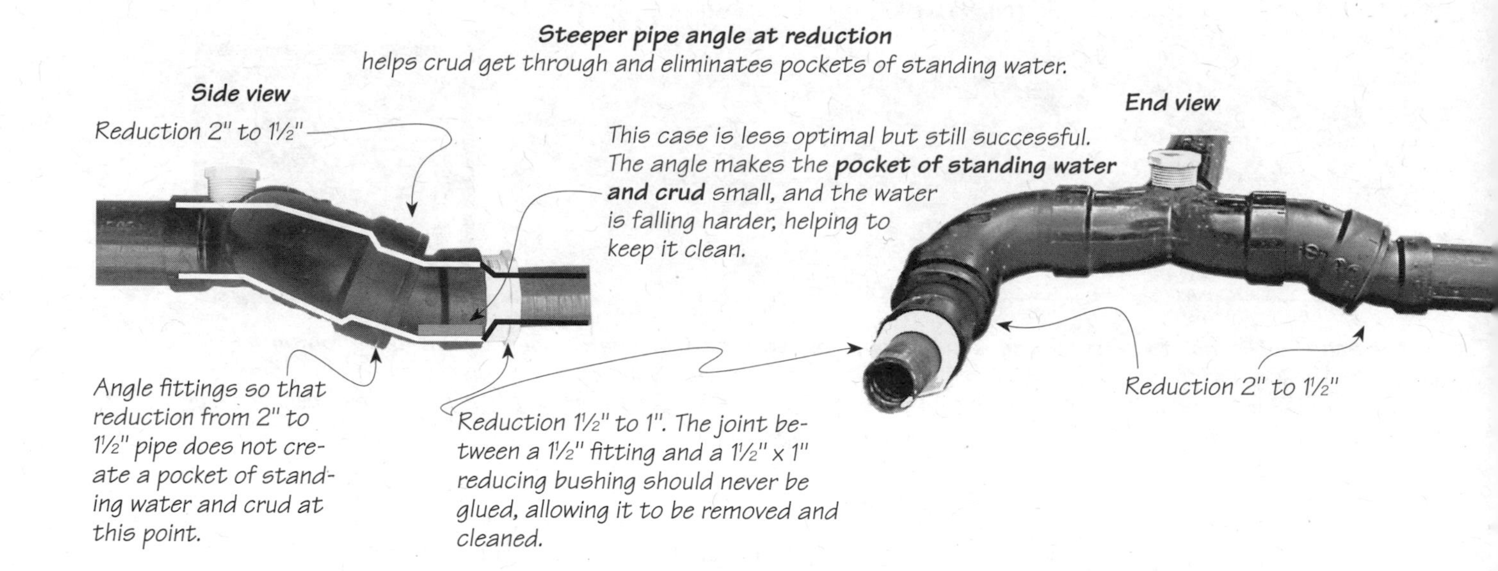

Steeper pipe angle at reduction
helps crud get through and eliminates pockets of standing water.
Side view
Reduction 2" to 1½"
This case is less optimal but still successful. The angle makes the **pocket of standing water and crud** small, and the water is falling harder, helping to keep it clean.
End view
Angle fittings so that reduction from 2" to 1½" pipe does not create a pocket of standing water and crud at this point.
Reduction 1½" to 1". The joint between a 1½" fitting and a 1½" x 1" reducing bushing should never be glued, allowing it to be removed and cleaned.
Reduction 2" to 1½"

Cleanouts, Inspection Access, and Rainwater Inlets

As with collection plumbing, you want to make it easy to inspect/unclog every part of the greywater distribution network.

A cleanout right after the line exits the house is convenient and can double as a means of bringing rainwater from downspouts into the system. After that, you need a cleanout every 135° of aggregate bend (see Provide Cleanouts and Inspection Access, Chapter 4). Three-way valves provide inspection access in all directions. A dipper box provides excellent access to every pipe connected to it. Long runs of pipe may require cleanouts at intervals, particularly if their slope deviates from ¼" per foot *(2%)*.

The flow splitter inspection ports and the outlets provide inspection and cleanout access to the distribution plumbing. Though the 1" *(2.5 cm)* openings are small, they have proven adequate so far (perhaps because virtually no clogs have been observed). The 1" access hole at the flow splitters is not big enough to run a snake through, at least not without ruining the plastic threads holding the access plug. Blasting water from a garden hose at the flow splitter access has been effective for unclogging.

If your cleanouts/inspection access aren't convenient for admitting rainwater, be sure to include some dedicated rainwater inlets. You can include a spur from the main greywater line to a convenient downspout. The best arrangement in most contexts is to have rainwater go elsewhere by default, but make it easy to manually reroute it to the greywater system just for flushing salts. In all but the driest climates it is inadvisable to plumb rainwater into a greywater system without an easy way of diverting it elsewhere. You don't want to overwhelm the system. Some greywater codes prohibit plumbing rainwater to a greywater system for this reason.

Branched Drain Outlet Design

The options for Branched Drain outlets present a trade-off: low cost, simplicity, and durability on one hand, hiddenness on the other. We'll consider them in order of simple to hidden.

Free Flow Outlets

From a free flow outlet, the greywater daylights for a few inches in midair after it flows out of the pipe and before it disappears into several inches of mulch (Figure 9.8a). The free

FIGURE 9.8: BRANCHED DRAIN OUTLET OPTIONS
(ELEVATION VIEW)

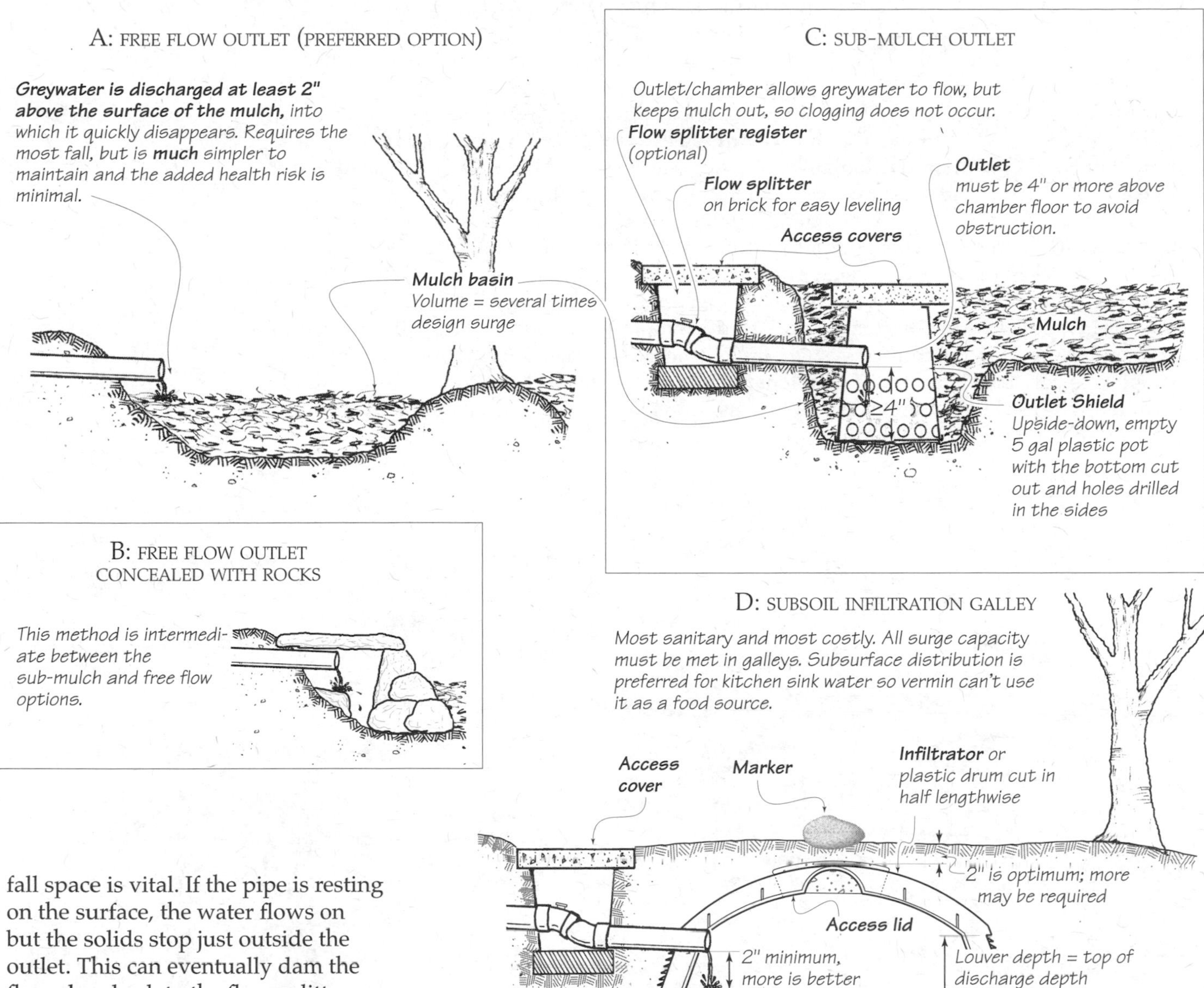
A: FREE FLOW OUTLET (PREFERRED OPTION)
Greywater is discharged at least 2" above the surface of the mulch, into which it quickly disappears. Requires the most fall, but is much simpler to maintain and the added health risk is minimal.
Mulch basin
Volume = several times design surge
B: FREE FLOW OUTLET CONCEALED WITH ROCKS
This method is intermediate between the sub-mulch and free flow options.
C: SUB-MULCH OUTLET
Outlet/chamber allows greywater to flow, but keeps mulch out, so clogging does not occur.
Flow splitter register (optional)
Flow splitter on brick for easy leveling
Access covers
Outlet must be 4" or more above chamber floor to avoid obstruction.
Mulch
>4"
Outlet Shield
Upside-down, empty 5 gal plastic pot with the bottom cut out and holes drilled in the sides
D: SUBSOIL INFILTRATION GALLEY
Most sanitary and most costly. All surge capacity must be met in galleys. Subsurface distribution is preferred for kitchen sink water so vermin can't use it as a food source.
Access cover
Marker
Infiltrator or plastic drum cut in half lengthwise
2" is optimum; more may be required
Access lid
2" minimum, more is better
Louver depth = top of discharge depth

fall space is vital. If the pipe is resting on the surface, the water flows on but the solids stop just outside the outlet. This can eventually dam the flow clear back to the flow splitter. Stuff falls away from free flow outlets rather than toward them. They never clog, they accommodate high surges, and they are easy to inspect—you'll note the functional status of the system just by spending time in your garden.

Free flow outlets require 3–10" *(7–25 cm)* more fall than the subsurface options, to create enough grade difference for the buried pipe to surface above the mulch. Leave the outlet visible or conceal it with a few large rocks.

Free flow outlets are allowed in Arizona, New Mexico, and Texas. In other areas, it takes either a visionary or inattentive inspector to approve them.

Sub-Mulch via Outlet Shield

Sub-mulch outlets emit greywater inside shields made from plastic pots, buckets, sawn-in-half drums, or infiltrators (see Figures 9.8c, 9.9, 9.10).

Free flow Branched Drain outlets *in a retaining wall.*

Water coming out
Flow splitters

These chambers create voids under the mulch where solids can fall clear of the outlet, preventing clogging. Sub*soil* chambers are sized to accommodate the entire peak surge, which then soaks slowly into soil. Sub-*mulch* outlet shields can be smaller, provided they are permeable enough that greywater can run out rapidly into the surrounding mulch basin, where most of the surge capacity is. A 5 gal flowerpot is the minimum size.

Mulch does degrade and must be replaced, usually annually, for the distribution to stay "subsurface." Don't use gravel if a gardener might ever live in the house. Gravel is the last thing you want in your garden, and a legally sized system would use a big truckload of it.

Nature abhors a void, so sub-mulch outlet shields tend to fill with crud—mulch and greywater solids—that needs to be mucked out. Outlet shields can trap small critters such as lizards, etc. One system owner suggests putting sticks in the shields as escape ladders.

Many codes require that greywater be distributed below the surface, but they don't specify *which* surface: soil or mulch. Mulch is preferable and has passed muster in permitted systems in many jurisdictions (e.g., Real World Example #1). Sub-mulch distribution is the best compromise for a permitted system. It yields improved sanitation at the cost of more complex design, installation, and maintenance. Go this route to satisfy code, or if you think vermin might use kitchen greywater outlets as soup kitchens. (If you have separate flows, consider making the kitchen outlets subsoil.)

FIGURE 9.9: HALF-BARREL OUTLET CHAMBER

FIGURE 9.10: FLOWERPOT OUTLET CHAMBER

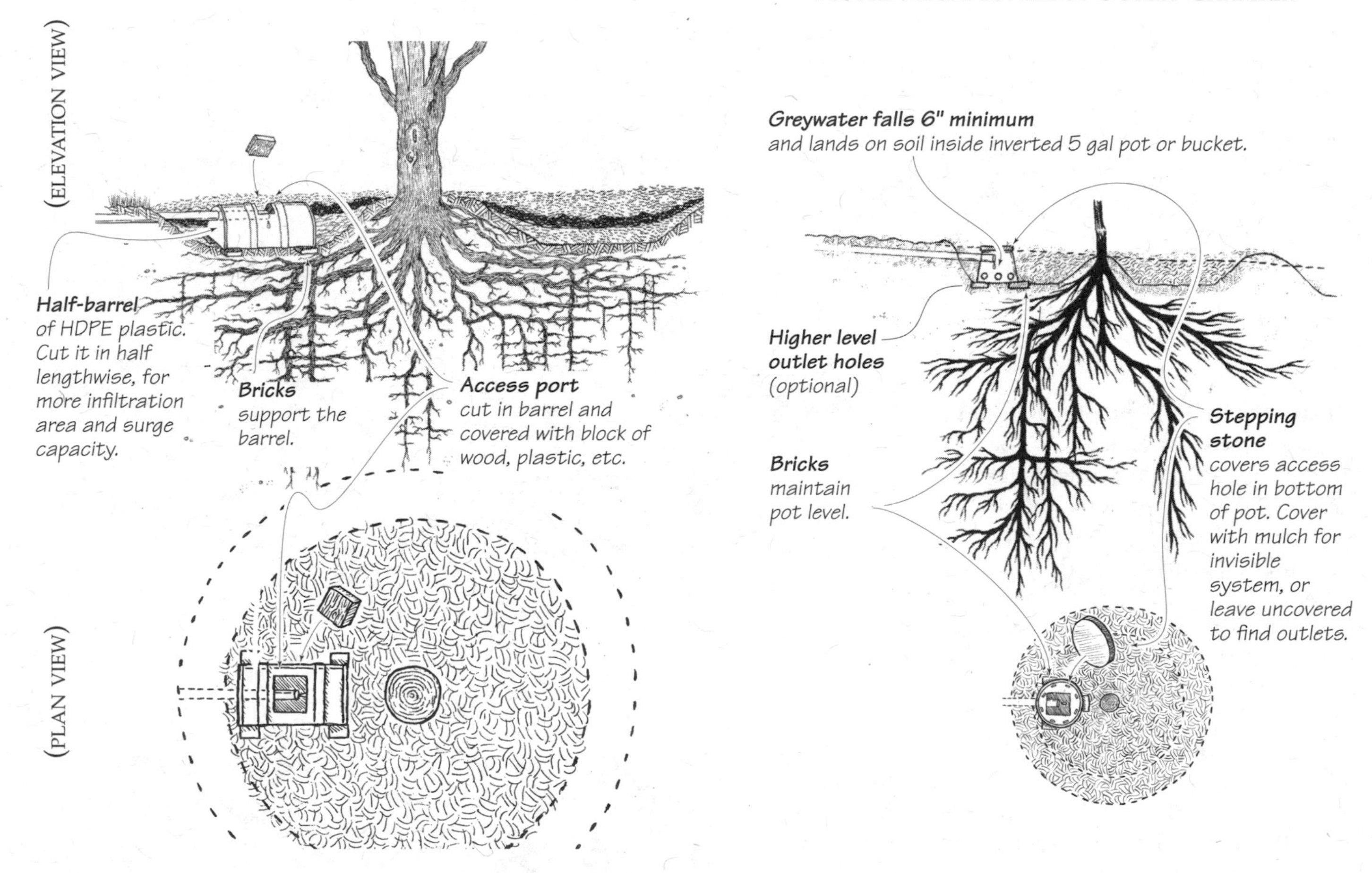
(ELEVATION VIEW)
(PLAN VIEW)
Half-barrel of HDPE plastic. Cut it in half lengthwise, for more infiltration area and surge capacity.
Bricks support the barrel.
Access port cut in barrel and covered with block of wood, plastic, etc.
Greywater falls 6" minimum and lands on soil inside inverted 5 gal pot or bucket.
Higher level outlet holes (optional)
Bricks maintain pot level.
Stepping stone covers access hole in bottom of pot. Cover with mulch for invisible system, or leave uncovered to find outlets.

Subsoil Infiltration Galleys

It is possible to design a system with completely subsoil distribution. In the previously mentioned options, the outlet is just the gateway to a large surge capacity and infiltration area in a mulch basin, which in turn can overflow safely under the mulch between basins. In the subsoil case, *all* the surge capacity and infiltration area is within the infiltrator itself (Figure 9.8d). Unless the surges are small and the perk rate high, you're going to need rather

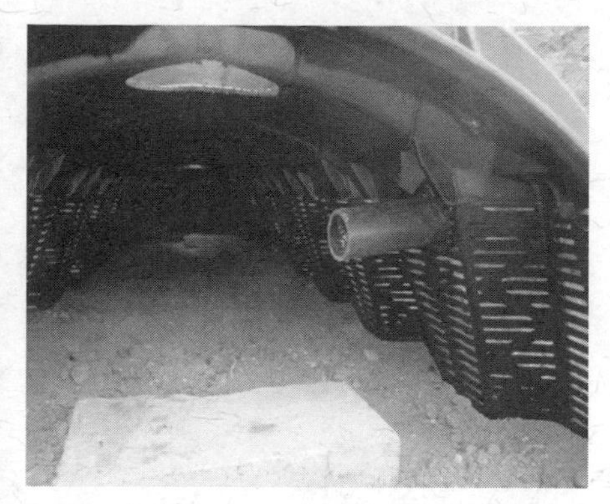

Infiltrator, *inside view.*

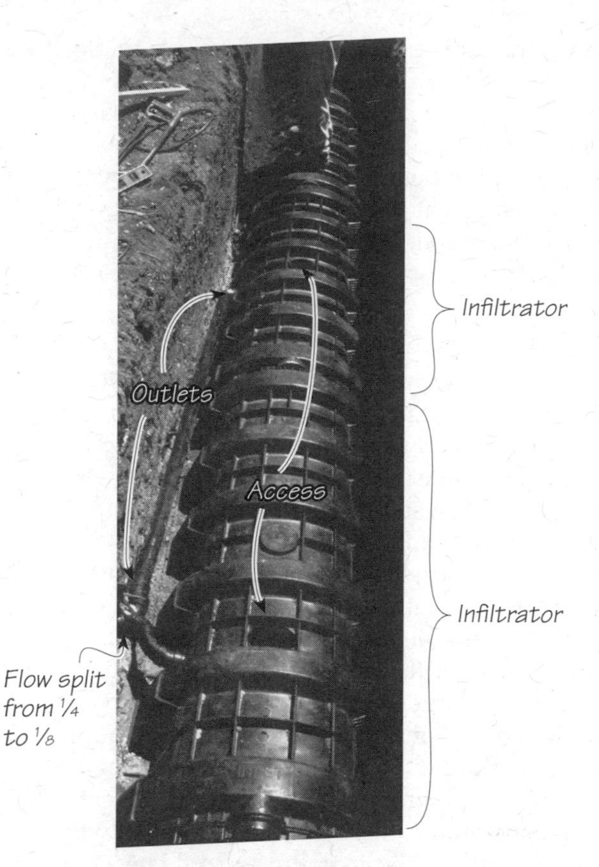

Infiltrators fed by a Branched Drain network. *Each of the eight infiltrators gets one outlet (not all are shown; two outlets visible here).*

large chambers to get adequate infiltration area (see Table 2.3, Disposal Loading Rates). It may be completely unfeasible for an inspected system. Not only is the code-specified area ultra-conservative, but the greywater quantity estimates are high. If you're lucky they might let you use the area of the overlying tree canopy and mulch basin as the treatment area, since it actually is. Infiltrators (see photos) are the easiest way to provide a lot of square feet of subsoil infiltration.[s13]

Subsoil Infiltration Galleys are so sanitary there's no health reason not to put blackwater into them. In fact, they may last longer due to the filtration by the septic tank. (See also Green Septic, Chapter 8, for more info and photos.)

Note: Infiltration Galleys are a proven technology for disposal of septic tank effluent, which is nearly free of clogging solids. Though promising, there is relatively little track record for application of raw, unfiltered greywater inside completely earth-covered chambers. Place them as high as possible in the soil profile (where perk and worm activity are greatest). Don't skimp on the size of the galleys. The success or failure of this design hinges mostly on distribution evenness and loading rate. Be sure to provide access for inspection/mucking out. Don't try a highly loaded subsoil galley without an alternative, backup infiltration area such as a mulch basin or mulch-covered area.

Outlet Positioning

Where exactly do you put these outlets? You want the tree roots to reach the water, but not get suffocated—see examples of outlet geometry at right. To avoid root suffocation from lots of greywater and/or low perk soil:

- Keep the outlets some distance from the trunk
- Provide multiple outlets per tree for large trees
- Plant the tree on a high island in the mulch basin (Figure 5.7)

Figure 9.11: Outlet Configurations

A: Young tree, sandy soil

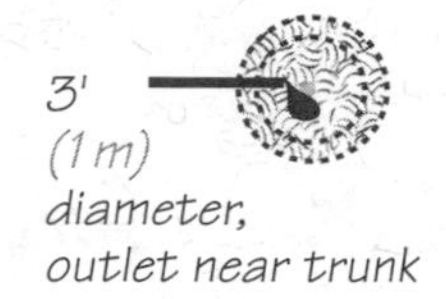

B: Medium size tree or young tree in clay soil

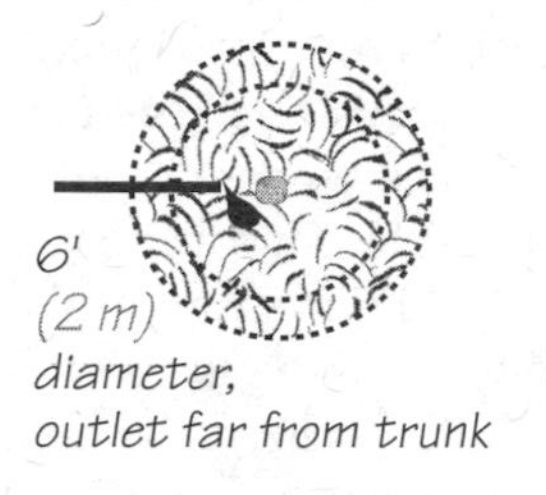

C: Large tree

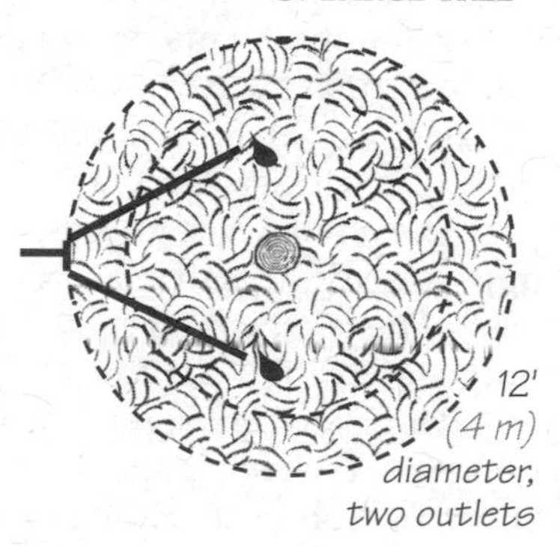

D: Very large tree

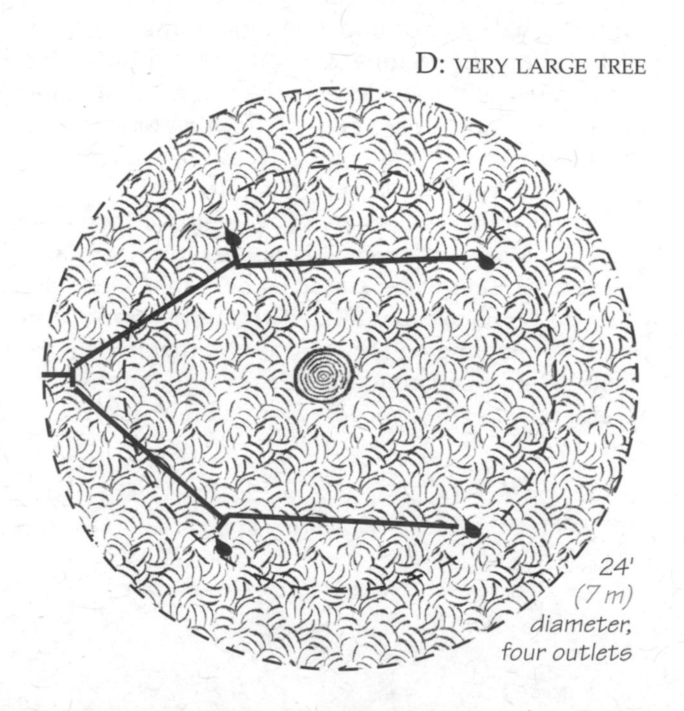

Except for a newly planted tree, the roots extend farther than you think: 50% beyond the canopy or more. If you have a young tree, arrange the pipes so the outlet geometry can be changed and / or greywater quantity increased as the tree, roots, and mulch basin grow. If your young tree needs supplemental irrigation until its roots grow out to the greywater source, burying an unfired clay pot next to the trunk is a great trick—you fill it with freshwater and the water slowly seeps into the soil. (It would clog if filled with greywater.)

Mulch Basin Surge Capacity

This section is included primarily to help you ease any fears your inspectors might have. In practice, surge capacity is rarely an issue unless the system is absurdly undersized and the overflow is to a problem area such as a public sidewalk. (For details, see Greywater System Calculations in our Builder's Greywater Guide.[6])

Branched Drain systems do not include a surge tank. All the surge capacity is in the pipes, outlet chambers, basins, and soil.

This is how you calculate the required surge capacity. First, determine the peak design surge using Table 4.2. For example, if a washing machine discharges while a bathtub is draining, this yields a surge of 75 gal in 10 minutes: 15 gal for the wash cycle, 15 gal for the rinse cycle, and 45 gal for the tub.[m] Four small mulch basins or one large one can hold 75 gal. The volume that counts is the volume below the basin walls or the outlet spill point, whichever is lower in height. When making this calculation, assume that the earthen basin walls will eventually wear down to half their original height.

If the outlets are free flow above the mulch, there is no impediment to the water quickly spreading in the basin. Still, the basin volume should be a few to several times the maximum expected surge if possible, to account for the volume of mulch, and for the walls wearing down.

If the greywater discharges into sub-mulch outlet chambers, surge handling is trickier. The chambers may drain slowly into the rest of the basin. The tendency is to drastically undersize the chambers. Beside low surge capacity, a small chamber has much less infiltration area, and tends to clog. The most conservative approach is to include enough chamber volume to accept the entire surge (see Figure 9.12).

Of course, for Subsoil Infiltration Galleys, all the surge must be accommodated in the chambers. (For most soils the short-term contribution of soil percolation to absorbing peak surges is not significant. Generally less than 10% of the surge can soak into the small area under the chamber in the few minutes the surge lasts.)

If the flow splitter preceding a chamber is higher than the top of the chamber, the entire chamber volume counts toward the surge capacity. If it is lower, only the volume below the split counts. This is because as soon as the chamber fills to the level of the splitter, that branch stops accepting water and all subsequent flow goes to the other branch. For certain geometries, this could result in nearly all of the excess flow going to the lowest chamber after the others fill, severely overflowing it.

The reliability and safety of a system designed conservatively, as described above, far exceeds that of the typical sewage treatment plant. Perspective: Greywater surfacing once in a blue moon is not a big deal when you consider that toilet water from failing septic leachfields across the country is pooling on the surface, and that in hard rain, municipal treatment plants abandon all pretense of treatment and discharge raw sewage into natural waters where people fish, swim, and extract drinking water.

FIGURE 9.12: OUTLET CHAMBER OPTIONS / SURGE CAPACITIES (PLAN VIEW)

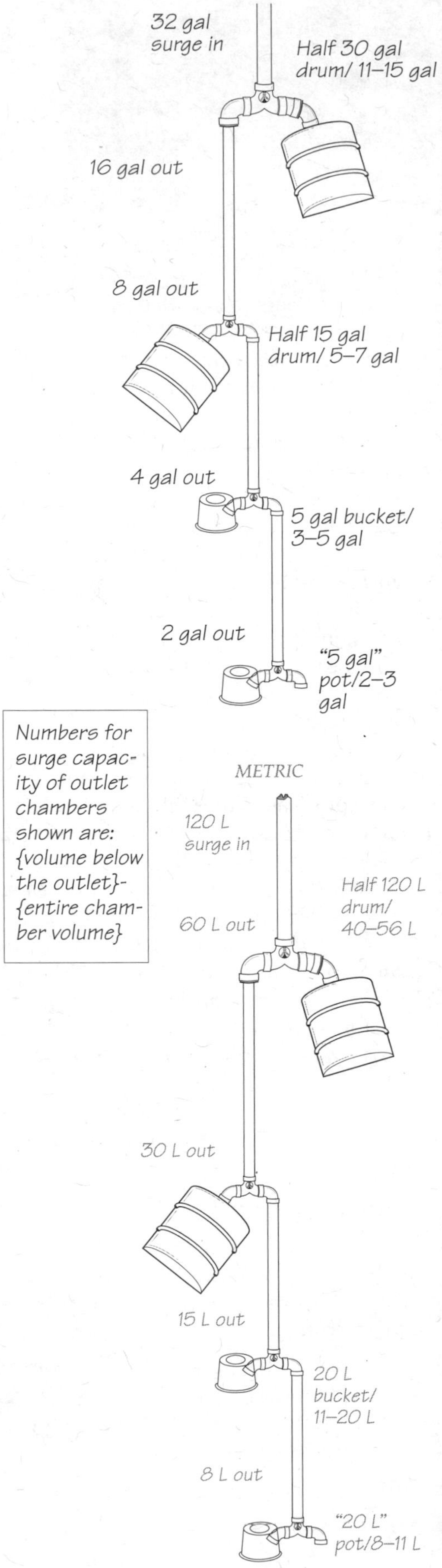

32 gal surge in
Half 30 gal drum/ 11–15 gal
16 gal out
8 gal out
Half 15 gal drum/ 5–7 gal
4 gal out
5 gal bucket/ 3–5 gal
2 gal out
"5 gal" pot/2–3 gal

Numbers for surge capacity of outlet chambers shown are: {volume below the outlet}-{entire chamber volume}

METRIC
120 L surge in
Half 120 L drum/ 40–56 L
60 L out
30 L out
15 L out
20 L bucket/ 11–20 L
8 L out
"20 L" pot/8–11 L

[m]***Metric:*** *284 L surge in 10 min: 57 L for wash cycle, 57 L for rinse cycle, 170 L for tub.*

Branched Drain Mental System Check-Out

If you are shooting from the hip you can skip this step. Otherwise, take a moment to look over your design and consider what will happen when/if:

- **Ten years go by**—Can the system be adapted to foreseeable plant growth and / or changes in the house's greywater production? Will you be able to locate buried parts of the system to check up on them? (See Map the System, Chapter 10.)
- **You sell the house**—Will the new owners be able to change the system to suit their habits and preferences? Will *they* be able to locate buried parts of the system to check up on them?
- **There is a very large, sustained flow of water from the house**—Will it all end up at the lowest basin when the others fill and back up to the flow splitters? Or will it overflow from all the basins? The latter is preferable.
- **The whole site floods with rain or rising water**—Conceivably, during a flood, a Branched Drain network could transport runoff water down, or floodwater up, from one part of the yard to another, where it could be harmful.

Taking these considerations to heart, adjust your design if necessary.

Branched Drain system installation *during a greywater workshop in Cottonwood, Arizona, with Brad Lancaster and Karen Taylor.*

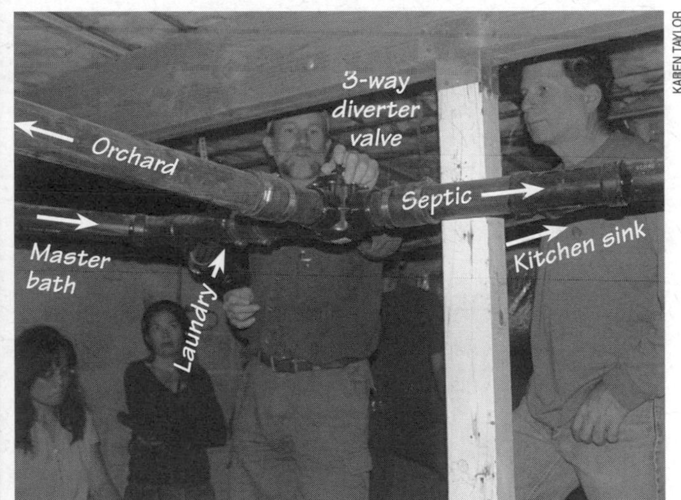

Three-way valve installed. *Wye on right is kitchen sink water joining the line downstream of the diverter. It may not look like it from this perspective, but all the pipes slope down 2% in the direction of flow.*

Chapter 10: Branched Drain Installation

Even if you're not installing a Branched Drain system, reading the detailed tips and tricks in this chapter will help you do better greywater collection plumbing and installations of other system types.

At this point, you should have your greywater and irrigation need connected on paper. Now it is time to connect them with pipe. Branched Drain systems are best built from top to bottom along the water flow, while checking ahead to confirm that there is enough fall left and that the outlets will come out in the right place at the right height.

To establish the starting point, you need to install your **collection plumbing** (covered in Chapter 4). Then install the Branched Drain **distribution plumbing** and **receiving landscape** as follows...

Double-Check Your Design

Obstacles such as concrete walkways, large roots, or buried pipes can block your way, forcing you to give up fall to go under them, or replumb the whole system higher to go over them. You may be able to burrow under a walkway by digging and/or boring with a garden hose (a nozzle at the end raises the pressure). You can cross over a walkway, especially next to a step, by snuggling the pipe in the corner of the step and covering it with cement. As a last resort, you can break a trench through a walkway, then patch it. A pipe can cross over a sunken dirt pathway in a long, gentle "speed bump" of dirt added on top. Some buried pipes can be raised or lowered to go over or under them at the height you want, by uncovering a long enough run that the obstructing pipe can be bent out of the way. This could possibly be done with a large root as well. Ditches can be bridged by a wooden plank with the pipe attached under it, or bridged by a larger metal pipe with the active pipe inside it. Make sure the pipes are out of the sun and harm's way and are not liable to settle or sag. Double-check the overall slope from beginning to end of the trench to establish that you can get where you want to without the outlets being too deep (see Measuring Elevation and Slope, Appendix B).

***A pad of stiff concrete mix** held in place by bricks awaits the dipper.*

Check for Buried Utilities

As with any digging project, check for underground utilities before digging. Greywater systems are generally not much trouble in this respect, because their lines typically run shallower than buried gas, water, electric, and phone lines, but it doesn't hurt to check. Often utility companies will come out and mark the location of buried utilities for free.

Dipper Installation

Skip this if you aren't using a dipper.

If there's a dipper box in your design, you want to install it early on. It consumes about 8" *(20 cm)* of fall—make sure there is enough. You want permanent access, as the dipper is a great place to access the system for flushing, snaking, changing irrigation zones, etc. Finally, you need good access to install it—it is heavy, and needs to end up exactly level (photos at right; there's also more on dippers in Chapter 9).

***Leveling a dipper:** Stand on it and wiggle it down into the concrete while checking levels in both directions.*

Once your spot is established, dig a few inches deeper than the dipper's height, compact the soil well, and make a pad for it out of concrete. Rebar or other reinforcement is unnecessary. Use a stiff (low water) mix and put temporary bricks or boards on the sides to keep the concrete from squirting out under the dipper's weight (upper photo). Position bricks so they take the impact when you drop the dipper in, then pull them out so you can settle it into place. The concrete should be about 1½" *(4 cm)* higher than the final dipper height, so the dipper is still about ½" *(1 cm)* higher after it initially squishes into the concrete. You can lower the dipper, but you can't raise it without taking it off and adding more concrete. Wiggle the dipper, tap on it, and if necessary jump on it so it slowly squishes down into place, checking the levels continually as you do so (lower photo). Remove the bricks and scoop away the concrete as it squishes out to allow the dipper to sink farther. Push it down by smaller and smaller increments as you get close—you don't want to screw it up now. You're done when it has sunk to the right height and is level from front to back and side to side. Don't touch it again until the next morning.

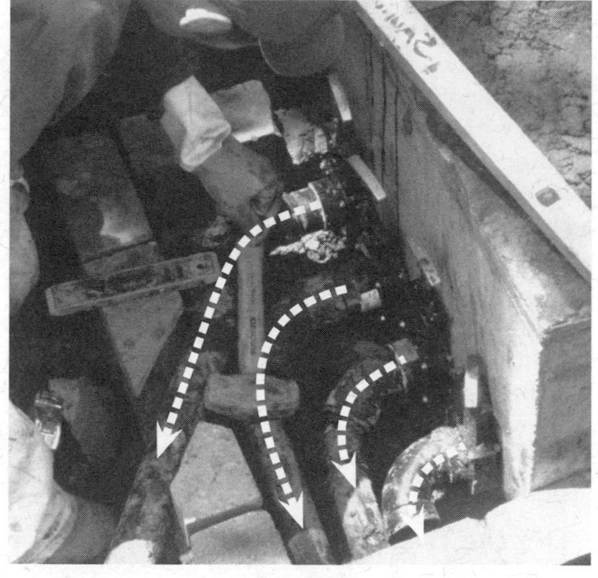

Dipper distribution box with a brick enclosure around individually valved outlets.

Connect Pipes and Fittings on the Surface without Glue

Make any adjustments needed and confirm that the layout is basically sound, the network flows from high to low, and the outlets are situated reasonably well. Lay the pipe on the surface without glue. You can cut the more certain lengths. De-burr the cut ends so crud doesn't hang up on them. Also, tip the pipes vertical and let any stones or dirt they've collected fall out before joining them.

Laying pipes on the ground without fittings to help visualize the layout. We moved them around until it seemed like the right trees would get the right amounts of water. I then drew the system, and laid out the pipes with fittings. Finally, we dug trenches, glued it all together, and buried it.

Dig Trenches

Dig the trench to the first flow splitter. How deep? Optimal is 9" *(23 cm)* and more than 1' *(30 cm)* is certainly not necessary. In most soils, this is deep enough that a truck could drive over without breaking the pipe. The greywater itself is released at this depth or shallower, so there is no health reason to dig deeper. In permitted installations you may be required to adhere to a minimum depth (8" is the minimum depth for greywater-to-drip-irrigation PVC lines in California).

In practice, it is often necessary to take the pipe all the way to the surface in problem spots (a low area on the way to a high area, for example). Considering that dumping raw greywater right on the surface is fairly common, it is hard to get too upset about greywater being *inside a pipe* at the surface. In a permitted installation you might get away with passing the plastic pipe through a steel pipe, using cast iron pipe, or covering the pipe with concrete. After all, these measures are allowed for pipes containing raw sewage.

The amount of work required to lay pipes depends greatly on their slope, as described next.

Marking trenches. Use string lines just to the side of the trench for straight runs, and powder (ashes, lime, whatever) over the actual pipe for curves. Straight lines are easy to map for posterity, but curved pipe is preferable to mid-run bends using couplings.

Laying Pipe with Plenty of Slope

Even when the ground slopes steeply, if your pipes *traverse* (cut across) the slope at any point they may end up with the bare minimum of slope to get where they need to. But if all your pipes have plenty of slope (more than 1" per foot, or 8%), you can throw the pipes and fittings in without glue, check the layout, spot check with a level to ensure they don't slope less than 2% anywhere, glue them, cover them, and you're done. You probably should include extra cleanouts in case the steeper slope results in solids getting left behind the water, *but you can skip all the extra steps below for Laying Pipe with Marginal Slope*—hallelujah!

Giant tree roots*—hopefully you'll be lucky enough to pass underneath without losing any fall, as we were able to do here.*

Laying Pipe with Marginal Slope

If your installation requires you to conserve fall, it is an involved, multi-step process to get bendy plastic pipe to slope precisely in dirt. You can never go *under* 2% slope. The less fall you have to work with, the less you can afford to go *over* 2% in any given portion. The smaller the range of acceptable slope becomes, the more fanatically you have to follow the steps below.

Dig the Trenches Slightly Deeper than the Pipe

Dig your trenches to a tolerance of +1", -0" *(+2.5 cm, -0 cm)*. It is easier to backfill under the pipe with sand or compacted earth where the trench is too deep than to dig dirt, rocks, and roots out from under the pipe in dozens of places where it isn't *quite* low enough.

Assemble Pipe and Fittings in the Trenches without Glue and Check Elevations

Assembling pipe *inside the trenches without glue, and checking levels.*

Level and assemble the system from top to bottom, occasionally jumping downstream to confirm that there is enough fall left to make it to the outlets with proper slope.

First assemble the plumbing as far as the first splitter without glue. Tip pipe segments vertical so any rocks and dirt fall out. Level plumbing coarsely to check that trench bottoms are deep enough everywhere to allow final levels to be achieved without more digging, and that you've chosen the right fittings and assembled them correctly. Note that pipe seats deeper and easier into couplings with glue than without; when they slide in all the way, the geometry changes slightly. Also, plastic pipe is almost always bowed. So it doesn't wreak havoc with your leveling, sight down the pipe and orient it so the bow is to the left or right, not up and down. Mark the up side of the pipe so you know which way to glue it.

Glue the Pipes Section by Section

Gluing is not strictly necessary. I've installed a couple of Branched Drain networks entirely "friction fit" (not glued). One joint in each system has come apart in five years. If you're unsure of what you're doing, or prone to changing your mind a lot, you may wish to leave part or all of the distribution plumbing unglued.

On the other hand, gluing helps ensure that proper levels will be maintained and the system won't fall apart as soil shifts and settles, roots sneak in, and plants grow. In reality, it is easy to change plastic pipe around, even after it is glued. Just saw into it and move it. I'd say the optimal method is to glue 90–95% of the system and leave a few strategic joints unglued. The things I most often leave unglued are:

- **Outlet extensions to trees**—When the plants have grown, pull the extensions out and the water discharges where the new feeder roots are, farther from the trunks.
- **Expensive fittings such as 3-way valves**—If there is pipe going into them, I often glue it (you can cut the pipe and still reuse the valve). But if there are street fittings coming out of the $50 valve, I glue with silicone sealer so the plumbing around the valve can be changed without having to chuck it. (See Build for Future Flexibility, Chapter 4, for other no-glue connection options.)

Load the glue applicator with just enough glue to cover the female coupling first and then the male coupling. Too much glue squirts out inside the pipe in gobs or strings, which harden and obstruct the flow. Once the glue goes on, you've got just a few seconds to get the fittings lined up right before it grabs. Drain fittings such as sanitary tees have a 2% downward angle built into them so properly sloped pipe slides straight in. This angle is visible if you look for it. For fittings that lack this angle, such as flow splitters and diverter valves, it is a nice touch to tweak the fitting or pipe as you glue it into the hub so it slopes downward a bit (photo at right).

On the whole, you want to glue the system from top to bottom. However, it is often convenient to glue subassemblies such as the one below, then glue them onto the end of the pipe and level them.

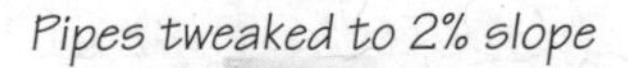

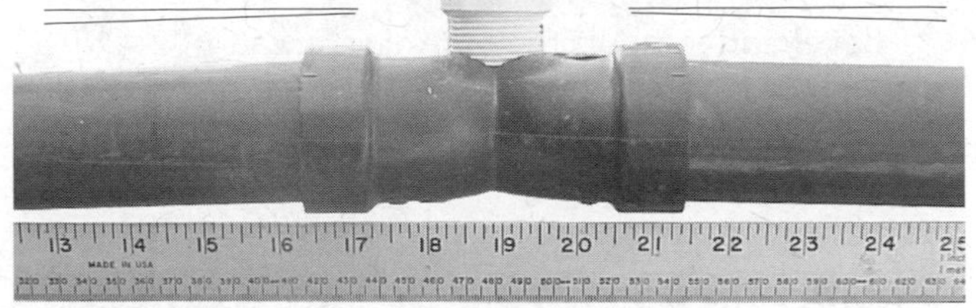

A 2% downward tweak on pipes *ensures smooth downward flow of water and solids.*

A mark shows how couplings line up—*a reference for gluing this subassembly correctly.*

Subassembly is then glued into the system as a unit.

You'll notice that some joints "matter": If they are not rotated correctly, the slope won't be right. Other joints don't "matter": The pipe can go in any rotation and the slope will still be right. Paying attention to this distinction can help smooth your work flow.

Make a mark to show how the couplings should be rotated to line up together. If the pipes and fittings are all pretty much in the same plane, you can rotate the fitting to be just barely past level on the side of the flow splitter (judged by the center lines from the molds, visible on the sides of the fittings). This ensures enough slope to facilitate flow from the pipe into the fitting. Once you've glued as far as a flow splitter, you can start friction fitting the next branch, or finish all the steps through burying the pipe before starting the next branch. The advantage of leaving the glued pipe leveled but uncovered is that you can easily change its level if necessary. The advantage of covering leveled pipe is that it holds still.

Sometimes you find the pipe is too low at the end where you're working. The solution is usually to measure slopes back up the pipe you've laid, find the stretch that's sloping the steepest, and flatten its slope. This enables all the downstream pipe to be raised in elevation. For example, changing 4' of pipe from ½" per foot to ¼" per foot slope gains 1" of elevation *(changing 1 m of pipe from 4% to 2% gets you 2.5 cm).*

Glued pipes ready for support/burial.

Level and Support the Pipe at Intervals

Level and support the pipe at intervals of 2–4' *(0.5–1 m)* or as needed. Sand is ideal for supporting and leveling pipe, but compacted dirt works too. Use dirt tightly wedged underneath to raise it, heavy objects placed on it to lower it, checking the levels continuously as you do so (photo at right). After this step, your pipe is like a bridge supported by pylons.

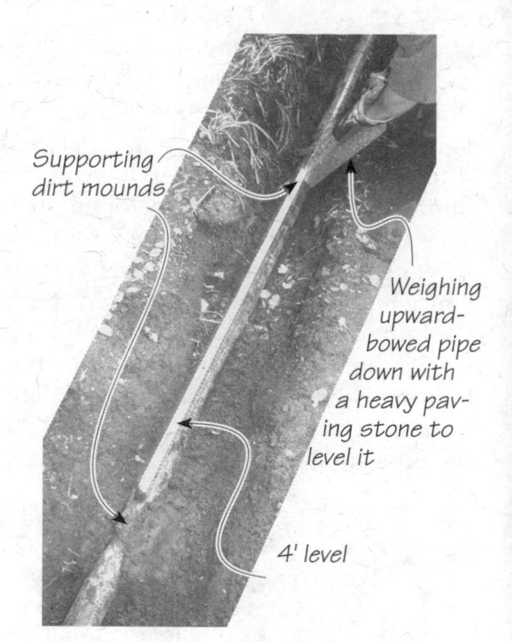

Pipe leveling
Pipe is supported at 4' *(1.3 m)* intervals with dirt mounds, pushed down to level as needed with weights.

Support the Pipe along Its Entire Length

Now pack dirt under the pipe along its entire length, between the pylons. Walking barefoot on either side of the pipe to squish the dirt in works well. Wet the dirt just right to facilitate compaction—wet dirt compacts much more, but if too wet it will squish out. Make sure this supporting dirt does not have leaves in it—when they decompose, the pipe will settle. Also make sure squishing the dirt underneath isn't raising the pipe, just supporting it. Check the levels of the pipe again. When the level is good and the bottom of the pipe completely supported with compact earth, fill in the trench to the top of the pipe. Leave the very top of the pipe clear. Wipe the top of the pipe clean to check the slope yet again (a pebble between the level and the pipe wrecks the measurement). Fix any place that's off. When you've done that, compact the earth again and recheck the levels one last time.

Bury the Pipe

At last you can bury the pipe. *Don't bury the flow splitters or valves yet.* Easy access is helpful for the system check-out later.

It doesn't matter if there are leaves in the fill on top of the pipe. Compact the backfill and leave it mounded up so when it settles it will come out even. Now trench the next section, and repeat the steps above.

Form Mulch Basins, Install Outlet Shields, and Plant Trees

This end of things is out of the range for most plumbers—it's landscaper's work. Once again, it's up to you to either do it right yourself or slightly bend your gardener's work conventions to meet special greywater requirements. Compared to regular mulch basins for fruit trees:

- Basins should be deeper
- Basins should be wider
- Walls should be compacted watertight by stomping or tamping
- Mulch should be a thicker layer of woodier (i.e., more durable) material
- Trees should be planted in taller islands

Mulch basin construction is especially critical for greywatered trees, as the trees will often get too much or too little water, due to the vagaries of greywater production.

It is generally best to wait until you have installed the pipe to finalize the *exact* position of the outlets and trees, because it's easier, for example, to move the outlet a bit than to turn 38° with a 45° fitting.

Once the pipes are in, it's time to construct the receiving landscape.

Dig Basins

Rake any leaf litter to the side and dig the mulch basins plenty deep. They only get shallower over time (review Mulch Basin Design, Chapter 5, for fine points; there are also mulch basin photos in Branched Drain Outlet Design, Chapter 9, and Real World Example #1).

Install Outlet Shields (if Any) and Plant Trees

Whether you use the 5 gal pots the trees came in, 5 gal buckets, half-drums, infiltrators, or something else for outlet chambers, a hole saw cuts very nice pipe passages. Thinner-walled chambers can be cut with a knife. Support the chambers in the mulch basins with stones or bricks so they don't sink into the wet ground. This is especially critical for chambers that contain both a flow splitter and an outlet. The flow splitter is supported by the pipe coming into its inlet and the pipe going out one side. The tendency is for the splitter to list toward the unsupported outlet side, which is just hanging in space in the outlet chamber.

System Check-Out

This is optional but fun. Offering to demonstrate that greywater does not surface is a card you can play with an inspector. Try it without the inspector first, to avoid surprises (see Builder's Greywater Guide*).*

Take the access plugs out of the flow splitters and blast a hose into each one from top to bottom to flush through any rocks that fell in during installation. Rocks are tough to flush with normal flows. Now try out the system at a low flow rate and confirm that the flow splits correctly. With a very low flow, the water might all go to one outlet. See which one it is. If it is a water-sensitive tree, consider fine-tuning the flow splitter levels so it goes somewhere else. Now run the hose full blast into a 5 gal *(20 L)* bucket, timing how long it takes to fill. Then run the hose full blast into a cleanout above all the splitters, timing how long it takes for the greywater to overflow a mulch basin. Multiply the minutes by the measured flow rate and you have the maximum surge capacity for your system. The measured surge capacity on installation day should be at least double the design's estimated surge, as actual surge capacity lessens between maintenance intervals.

For the inspector demo, you could fill your bathtub and sinks to the brim with stoppers in place, fill the washer, turn it off just before it pumps the water out, then turn on the shower full blast and let everything rip. *Voila*: No daylighting, no problem!

Map the System

You may be tempted to skip this step, but if you're like most people, the day after you've buried the system and raked the leaves back, you'd be hard pressed to locate the pipes.

The perfect cover for outlet chambers and valve boxes is a flat stepping stone, which hides their inner workings while *marking their location*. In case these markers get moved or buried, it is a good idea to record some measurements from stable reference points (buildings, large trees, concrete walkways, big rocks, whatever you think isn't going to move in the coming decades).

To map the system, note distances to each important point—flow splitters, registers, outlets, diverter valves, cleanouts, and dipper box—from two stable reference points on your biggest-scale site map. Ideally these points form a right triangle with the thing you're locating. Using three points is more work but improves mapping accuracy (see Figure 10.1).

Take snapshots of every part of the system. Number them and then note them by number on the site map where you were standing when you took them, with an arrow for the direction you were facing (Figure 10.1, and Real World Example #3). Put this site map and the photos in your house files. Low maintenance has a flip side: Because the system will work without you even looking at it for years at a time, it is very easy to forget where the parts are.

To find these buried treasures, hook your tape measure on the reference point (the corner of the house, say), then scribe an arc on the ground at the recorded distance. Now move the tape to the next reference point, and scribe an arc from there. Where the arcs cross, you'll find your treasure.

FIGURE 10.1: LOCATING SYSTEM COMPONENTS ON A MAP

Measurements from two stable reference points at approximately right angles can be used to locate these buried flow splitters in the future.

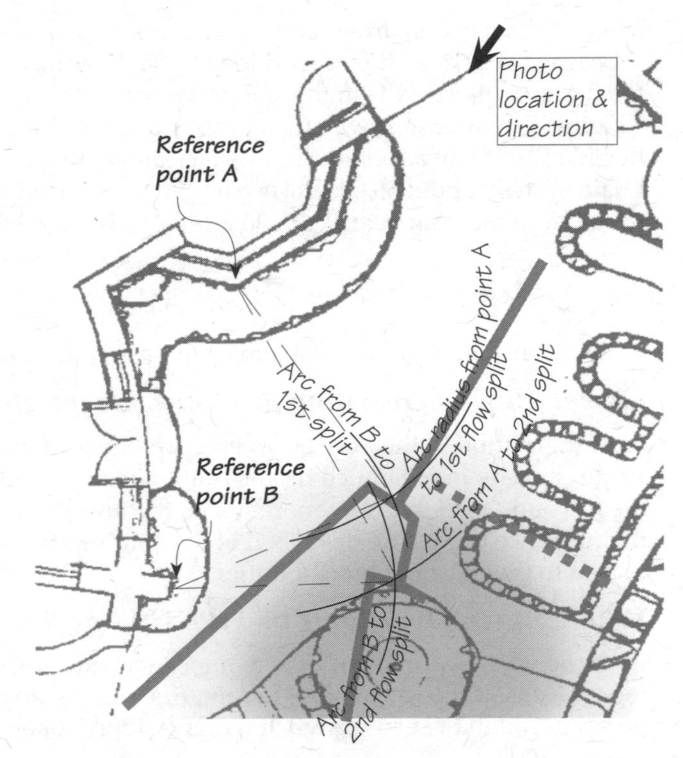

Cover Up

Now you can cover the mapped flow splitters and outlets with dirt and/or stepping stones, and fill the mulch basins with mulch.

Branched Drain Maintenance

These systems have held up very well for the years they've been in existence (since 1998). They are so dead simple it is likely that they will not be affected much by the passage of a few more years, or possibly decades. The required maintenance is minimal.

Service Mulch Basins

Mulch basins need fertilizer and new mulch annually. When that's all they need, just toss it on top. Each few years, the basins need to be emptied of mulch and reformed, larger if the trees are growing. You may also wish to extend outlets or add more splits. Stomp in the roofs of any burrowing animal holes, as these could drain water out of the basin and cause erosion. Refill the basins with fertilizer and mulch, putting clippings and yard waste on the bottom and the most attractive mulch on top. This normal garden maintenance is even more important for greywatered trees, as it keeps the greywater covered.

Clear Outlets

Check the outlets. Are the outlet shields (if any) filling with dirt or getting buried so the water can't get out? Remove the dirt with a rubber-gloved hand and trowel. "Rough mulch" (twiggy, woody clippings or chips) around the chamber enables free passage of water for much longer than materials like leaves, straw, or grass clippings. Don't replace the covers on the chambers until after the flow splitting test, below.

Check Flow Splitting

Turn on an inside fixture to feed the system, or stick a garden hose in a cleanout, and confirm that water comes out all the outlets and splits more or less as it should. (It takes surprisingly long for the water to reach the outlets—longer than it takes to walk to them after turning the tap on.) Are the outlets that are supposed to discharge, say, ⅛ of the flow, all flowing at about the same rate? With lower flow, the test is more stringent. If the flow seems uneven, flush the system by opening the flow splitter access plugs and blasting a garden hose into each split. Now put the plugs back in and check again. If the splitting is still not correct, check each splitter in turn to isolate the problem. If there is a clog that can't be flushed from a splitter's inspection access, blast water up from an outlet, using a rag or adapters to make a semi-pressure-tight connection to the garden hose. Once the blockage is dislodged, flushing through the accesses should be sufficient to clean out the system.

Flush Salts

In a rainy climate with free-draining soil, this step is unnecessary.

Once a year or so it is a good idea to flush the system with copious freshwater or rain to flush accumulated salts in the soil down below the root zone (see Preserving Soil Quality, Chapter 5). The easiest way to get rainwater into the greywater system is via a hose or flexible PVC from a roof gutter into a cleanout above all the splits. In one existing Branched Drain system, about 50,000 gal/year *(190 m^3)* of clean rain runoff from an uphill driveway is routed into the system at about 30 gpm *(110 lpm)*.

Branched Drain Troubleshooting

Here are some known problems, with suggested solutions:

Outlet Shields Overflow or Won't Accept More Water

If this happens, the flow between chambers and mulch basins is too restricted, or the surge capacity in the mulch basins is not adequate. With some layouts, if the upstream outlets are full and won't take any more water, the lowest outlet is overwhelmed. Try the outlet maintenance procedure described above, but you probably need to put in bigger shields or switch to the free flow type of outlet.

Whole Basin Overflows or Fills Up Past Outlet Level in Rainy Season

If the whole basin overflows, you need a wider, deeper basin, or to divide the flow among more basins. If this occurs due to rain, first make sure no runoff is getting in. If the problem persists, you may need to switch to long, sloped, open ended basins (described in Real World Example #4).

Animal Burrows Drain Water from Basins

Although this has not been problematic to my knowledge, gopher or mole tunnels could transport greywater out the bottom of the basin and off to who knows where.

The potential for this problem is greatly reduced by good design and installation. With a flow split in eight or more, then slowed and absorbed by mulch, the flow rate is so low that even if some water does escape through a gopher tunnel, it is unlikely to cause erosion or even get out of the tree's root zone before it soaks in.

If burrowing-animal-induced drainage would be disastrous, the entire basin could conceivably be armored above and below with ½" wire mesh before planting.

Flow Does Not Come Out One or More Outlets

It may just be a leaf or a piece of lettuce forming a temporary dam. Such blockages resolve themselves on their own. Another possibility (in an unglued system) is that a friction fit joint has come loose.

Check Flow Splitting as described on previous page. If blasting downstream with a hose doesn't work, stuff the hose in the end with a rag wrapped around it to tighten the connection and blast upstream from below. If that works, part of the system may have shifted out of level, so the flow is not going to one side. The only solution is to dig the blocked section up, re-level it carefully, and backfill around it carefully.

Outlet Chambers Become Slug Nurseries

Most systems have a few resident slugs. One had 30–50 in each outlet chamber. The owner of this system pointed out that intentionally creating slug lounges is a known organic control technique; you can go to the lounge and wipe them out, if you so desire. Pouring a huge pot of boiling water down the pipes might be a low-effort way to knock them back.

Grease

Pouring the occasional pot of boiling water through the system is also effective to prevent grease buildup. The grease melts and moves on to congeal on the surface of the soil. If you expect a lot of grease, upsize the infiltration area, or install a grease trap. A grease trap is a *small* inline tank in which grease cools, coagulates, and floats to the top. They are available commercially or you can make your own. They tend to get gross and stinky, especially if too big or if water sits in them over 24 hours.

Foul Smell

One of the big attractions of a Branched Drain system is the near-total absence of anaerobic conditions when the system is designed and constructed properly. This is because every part of the system slopes down, so it drains completely each time water comes in and there is no pooled water to turn anaerobic—well, almost none. There is a bit of water in each fixture trap, and if the system incorporates a dipper box there is up to 1½ gal *(6 L)* in the tray. If no water goes through for 24 hours or more, this water can start to stink. The stink will go out through the vent system to the roof, so you won't notice it until you use some new water, which pushes the old stinky water out to the yard. Any odor thus generated should be minor and transient.

If you have a persistent foul smell, it is probably due to a temporary dam of crud creating a long wedge-shaped lake inside a pipe; incorrect levels in the pipes, creating anaerobic U-shaped lakes; or an undersized, waterlogged infiltration area. Here's what to do:

- **To fix crud dams**—Do the flushing procedure under Branched Drain Maintenance, above.
- **To fix U-shaped pipes**—Dig up the U-shaped sections and re-level them.
- **If the infiltration area is too small for the amount of water it is receiving**—Use less water, make the infiltration area bigger, add provision to alternate between irrigation zones, add another zone, or divert greywater to another disposal option. Kitchen sink water is the most effective to divert, as the compost in it sucks a lot of oxygen from the soil.

Branched Drain Variations, Improvements, and Unknowns

Caution: Try at your own risk. None of these designs are proven....

Supplying Box Troughs in a Solar Greenhouse

The many reasons to send greywater to a greenhouse in a cold climate are summarized in Solar Greywater Greenhouse, Chapter 8. Greywater has been supplied to 3'-wide raised beds through box troughs. The unproven variation—which could prove simpler and easier to maintain—is to supply box troughs by gravity through a Branched Drain network instead of with a flooding dose from a Drum with Effluent Pump or a dosing siphon.

Green Septic System

A Branched Drain can distribute clarified septic effluent to Subsoil Infiltration Galleys. See Green Septic, Chapter 8.

Automatic Alternation Between Irrigation Zones

Automated switching between irrigation zones gives the soil a chance to rest, without someone having to remember to switch a valve manually. Manual switching tends to happen sporadically until the novelty wears off, and then not at all—certainly not at the optimum frequency of twice a week or so.

Automated switching could be done with a timer-controlled motor drive on a Jandy 3-way valve. (This is a common configuration for this valve in its intended use: pool and spa installations.[s2]) Similarly controlled valves could send a portion of the greywater to the septic/sewer when only some of it is needed. Or, admit freshwater when the amount of greywater is insufficient, or send water to different places based on readings from moisture sensors in various zones… The more you delve into it, the more possibilities you'll dream up. (But, before embarking too far on this course, see Error: Overly Complex, Delicate, and/or Expensive System, Chapter 11.) Perhaps someday someone will think of a simple, non-electric solution for switching greywater automatically between two zones for decades without intervention, like the rest of the system.

Supplying Bogs

Branched Drain networks have not been used to supply wetlands or bogs to my knowledge, so there could be some unforeseen difficulty. However, the means of getting the water to a wetland should have little effect on the proven ability of wetland plants to treat water. Clogging at the inlet is the most likely issue. Clogging by unfiltered greywater has successfully been avoided in wetlands by using a 3' x 6' *(1 m × 2 m)* infiltrator surrounded below and to the sides by rocks a few inches in diameter, and above by gravel.

The multiple outlet feature of Branched Drain networks serves little purpose when supplying one pond. The water attains the same level and extent whether it has one inlet or 16. I could see Branched Drains supplying a sort of self-equilibrating bog, which might occasionally be a pond but is otherwise just really wet soil. Multiple inlets would distribute the water over a wider area, so there would be standing water for shorter periods of time or not at all. The bog would equilibrate by plant growth or death as the water needs of the plants varied. As long as there was any greywater at all, the plants around the outlets would live to recolonize the bog when the balance shifted in their favor.

Be sure the bog does not intercept surface runoff, or it could easily be overwhelmed.

More than Four Levels of Splitting

More than four levels of splits (i.e., 16 outlets from one inlet) might work well using double ells; however, this hasn't been tested yet. The difficulty is that the water flow gets too low to push individual crud particles along (see Figure 4.7). A dipper or dosing siphon, which introduces peaks into the flow, would help a great deal to achieve finer splitting.

Smaller Pipes or Other Types of Pipes

It is preferable to use smaller and smaller sized pipes as the flow is halved and then halved again. This saves plastic and improves flow (Figure 4.7, again). It might be possible to make the finest branches with ½" class 200 PVC pipe (so far, the smallest I've heard of being used was 1" PVC).

Also, it may be possible to make Branched Drain networks out of welded HDPE (see Radical Plumbing, end of Chapter 4).

Double ells, 3-way valves, dipper boxes—none of the most important components of Branched Drain networks are specifically designed to be used the way they are in this system. If you own a factory for making plumbing fittings, contact me and we'll get some more optimally designed fittings made.

Very Hard-Freeze Climate

This will probably work, but to my knowledge, a Branched Drain has not been tried outdoors in a climate colder than high altitude in New Mexico, where temperatures drop to -10°F *(-23°C)*. As every part of the system drains completely, there are no pools of water to freeze. Greywater is probably warm enough to melt a film of ice inside the buried pipes with each use, so ice buildup would not occur.

Free flow outlets might get blocked by ice buildup, as drips freeze at the outlet. Oversized outlet chambers under a thick, insulating blanket of mulch and snow are oases of microbial activity warmed by the heat in greywater. These should work in all but the most extreme climates, in which case a Solar Greywater Greenhouse is indicated. (See also Appendix C, Cold Climate Adaptations.)

Very Low Perk Soil

The logical adaptations are large-capacity outlet chambers, free flow outlets, oversized mulch basins, and trees planted on high islands (Figure 5.7). It would probably be necessary to have two or more alternating infiltration zones, so they could each rest and have aerobic conditions restored. If someone dreamed up a simple system that automatically switched zones at intervals of time or after a specified amount of water was delivered, that would help.

Grease-Laden Water

We don't have enough information about the result of hard-core carnivores running their kitchen sink water through a Branched Drain network. It may be that grease will congeal inside, giving the pipes arterial sclerosis (narrowing of the pipes) as occurs in regular collection plumbing (and blood vessels). Grease may also clump in such a way that the flows split unevenly or not at all. If you expect a lot of grease, follow the recommendations mentioned in this chapter under Branched Drain Troubleshooting/Grease.

Cont. from p. 76

Chapter 11: Common Greywater Errors

Every reader should at least scan this chapter for applicable pearls of wisdom.
Don't, however, let it scare you. Even the most pathetically misguided greywater systems rarely cause actual harm, and for the few that do, it's not much (there has not been a single documented case of greywater-transmitted illness in the US). These pitfalls are easy to avoid, and even if you fall into one, odds are your system will still show a net benefit relative to the alternatives.

Although our readers do considerably better than average, they are not totally immune to the siren song of some particularly appealing greywater mistakes. Around 1995 we implemented a two-pronged approach: warning people about mistakes as well as steering them toward preferred practices.

Over time, we've accumulated quite a compendium of common errors. A few of the issues:

- Many complex greywater reuse systems are abandoned, and many simple ones achieve less than 10% irrigation efficiency within five years.
- Some greywater systems consume a lot of energy and materials to save only a little water.
- Economic payback time for some overly complex systems is longer than the system's life.
- Claims made for packaged systems are often inflated.
- The majority of "successful" (i.e., give more than they take) systems are so simple as to be beneath recognition by regulators, manufacturers, consultants, and salespeople.
- Hundreds of websites actually illustrate and promote the errors that follow. Some of their content is reproduced from manuals of the early 1970s, despite having been discredited years ago.
- Greywater best management practices remain under-publicized.

Of the greywater systems in the US, probably 15% are achieving most of the benefit they should, 80%+ could easily do better, and a few have overall negative net benefit.

Pondering the failings led our design path further and further from mainstream greywater thought. The more realistic we get, the lower we aim—and the more often we hit the target. Our designs hold up well in the field. In more than a dozen book revisions we've had to take back very little we've said.

The following collection of mistakes took many intelligent people many years to make, recognize, and then fix. Each error is followed by the preferred practice. We update this compendium constantly and post it to the web, as a public service (oasisdesign.net/greywater/misinfo). We hope you benefit from it. To help others benefit from your perspectives and experience, please share them with us, especially if you've made a mistake that isn't in our collection. We'll post them to our website and include them in future book revisions. If you've got an outstanding mistake, we'll send you a free book.

Error: Assuming It's Simple

Most of the errors that follow stem from this fundamental error, which in turn stems from:

1. Recognition that in most cases the resource potential of the greywater (water, nutrients, embodied heat) is not worth very much in the scheme of things, and that the costs of poor management (minor health threat, smells, etc.) are not very high either.
2. Failure to realize that a greywater system that achieves common goals (e.g., saving water) is a more site-specific and user-specific design issue than almost any other green home technology—more than solar heat, composting toilets, ecological building materials, etc.

In other words, greywater seems like it *should* be simple, so people often don't allocate greywater system design the effort required to achieve the performance they expect.

Preferred practices: Either lower your goals or put more thought/energy/money into the design. Reading this book is a great start. If your site is difficult, your goals high, or your time short, consider a design consultation as well (oasisdesign.net/design/consult).

Exceptions: If the performance goals are low enough for the context, greywater systems *are* simple. The Drain Out Back is a very common example of a "system" that doesn't require much effort at all and still offers advantages.

Error: Out of Context Design

Constructed Wetlands in the desert. Irrigation of a swamp. Sand filtration, ozonation, and pumping uphill for flushing toilets in a residence. These are valid designs but applied in the wrong contexts. In a culture where standardized solutions are the norm, we must remind ourselves constantly to pay primary attention to the context. Once the context is seen clearly, the steps to deal with it come out of the mist.

Preferred practices: All appropriate technologies are context specific. However, greywater reuse is *extremely* context specific. ***The one universally applicable principle for all greywater system design is that there are no universally applicable principles***. Many variables have the potential to change the design completely. Carefully check the list of design variables (Appendix A). General background on context specific design can be found in our book *Principles of Ecological Design*, from which the sidebar at right is excerpted.

Exceptions: There are no exceptions to the "no universal principles" rule.

Context Specific Design

While there are no solutions that apply universally, there are a variety of approaches and patterns that can be applied to generate the optimum solution for any need in any context.

This is a vitally important principle of ecological design. In all cases the greatest efficiency—and quality as well—is achieved when the power of the tool is well matched to the task at hand. Frequent technological overkill is one of the saddest sources of waste in our society. Elimination of overkill does not involve real sacrifice. The resources saved by using simple tools for easy tasks can be applied toward performing more difficult tasks. Using transportation as an example, walking would be used for the tasks for which it is adequate, bicycles for distances too long to walk, busses and trains for distances too long to bike or in bad weather, planes for speed or great distances, and cars for special applications such as ambulances, mobile homes, or workshops. Among countless other benefits, getting superfluous cars off the road would enable the necessary ones to get around without being choked by traffic.

Cleverly matching the power of the tool to the task at hand is cheaper, healthier, has lower environmental impact, and is more enjoyable—but ultimately more powerful than any single solution. Though uniform solutions appeal to centralized bureaucracies, attempting to implement a single solution across the board, without regard to context, will generate a host of new problems. Bare sufficiency produces optimal growth; deficiency is stunting, excess is unbalancing.

Those readers who are used to single solutions will be quick to point out situations where, for example, a composting toilet is unsuitable. No solution is applicable across the board, few technologies should be eliminated completely. What is suggested is that a range of solutions be matched to the range of contexts using common sense.

—From Principles of Ecological Design[15]

Error: Overly Complex, Delicate, and/or Expensive System

These systems miss the big picture and result in negative net benefit. A typical residential greywater system saves $60 worth of freshwater a month. If the system costs more than several thousand dollars it is probably overbuilt. The owner and the Earth would be better off wasting the water than the pumps, valves, piping, filters, and electricity.

Achieving "actual net benefit" can be a challenge. Systems are often built anyway, with vague allusions to a "demonstration project." This justification should be used sparingly, otherwise we'll end up with lots of demonstrations of resource waste.

In a residential context, any system that uses a pump and/or filter or costs more than you spend on water in several years is suspect. Disinfection is extremely suspect. Systems that entail massive, permanent disturbance to the planted area (such as the State of California's Mini-Leachfield system, which involves burying truckloads of gravel in the garden) are also missing the point.

Unfortunately, some greywater laws (notably the CPC/UPC) drastically reduce or completely destroy any possible economic, ecological, or community benefits by mandating wasteful design.

Preferred practices: *Choose the simplest possible design, and build it as well as you possibly can.* Keep one eye on your original goals and the big picture throughout. Ask yourself what the system is likely to look like in flood, in drought, or in 20 years after four owners. What is most likely to fail or be abandoned? How is the system likely to be "patched"? Often, the way systems get modified later to comply with reality is the way they should have been built in the first place. If no future "patch" seems acceptable, chances are the system should not be built that way at all.

Exceptions: New construction economics can justify much more complex systems.

Drought emergency causes greywater to skyrocket in value. If greywater is the only way to save a $20,000 landscape during a drought, an expensive system that falls apart in a year may be justifiable. On the other hand, a very well built, simple system that lasts for decades could possibly be made for the same price.

Chronic water shortage has the same effect. I designed a greywater system for a house where the only water supply is rain, stored for the eight-month dry season in giant cisterns (Real World Example #3). The greywater system cost a few thousand dollars, but provides the same amount of water as a $10,000 cistern, uses a fraction of the material, and solves the problem of greywater disposal.

Acute disposal problems change the picture as well. At one Alaskan oil camp, the greywater from several hundred people was boiled down to ash, and the drums of ash shipped to a toxic waste dump in Texas. Compared to this, *any* system would be cheaper, simpler, and more ecological.

The owners of a hotel in Big Sur were sending out the laundry at a cost of a few thousand dollars a month because they could not make a conventional onsite disposal system that could handle the water. A large, complicated, but well-built greywater system enabled their staff to do the laundry, and the hotel paid off the investment in less than a year.

System spotted at the Los Angeles Eco Expo that demonstrates a number of mistakes.
First, there's overkill. For several thousand dollars it does what a $500 Drum with Effluent Pump does. Second, the pictures on the poster show filtered but unsterilized greywater coming out at high pressure through sprinklers and inundating a concrete walkway from both sides. That violates three health principles (at least it didn't show kids dashing through the sprinklers). The tank is also too big, so that violates a fourth...probably a record for one system.

Well-designed, proven systems that last decades can pay for themselves even if complex. The US EPA State Revolving Fund program qualified the ReWater system[s5] for loans that had previously only been available to cities for conventional centralized sewer systems, primarily because the system lasts long enough to pay for itself.

Sheer volume shifts the economics drastically. Institutions with several thousand gallons a day of greywater production and a similar amount of irrigation need often find that a complex treatment system enabling efficient distribution can be paid off quickly.

Regulatory necessity. Regulators are generally more comfortable with complex, expensive, engineered systems than with biological systems that rely on the mysterious workings of nature. It may be that an overly complex greywater system is the best alternative to an even more overbuilt conventional system, if that's all they'll let you do.

Error: Mansion with a Greywater System

This is a specific case of the previous error. A greywater system for a large house with acres of irrigated area and just a handful of inhabitants is more likely to have negative net benefit. In this situation, the value of the greywater is literally a drop in the bucket compared to all the other waste going on—and attempting to capture it just adds more waste, in the form of hundreds of feet of extra plumbing. When you add the level of construction perfection and use convenience required to fit in with the rest of the house, the resource cost of the system compared to its use value can become ludicrously out of balance.

Preferred practices: *Build a smaller house.* A house half as big built with completely conventional materials and systems is almost certainly more ecological than a "green" mansion. Focus money and energy on a good, space-efficient design, and you'll have lots left over for state-of-the-art green materials and systems.[29]

If you must have a monstrous house, don't dual plumb it—it will cost too much plastic and money. *Reuse combined greywater and blackwater*, in a system that can handle it (see the System Selection Chart). You drastically reduce plumbing cost, get significantly more water and nutrient volume, and in systems that use drip irrigation, get much higher reuse efficiency.

Exceptions: None, really.

It is frequently argued that giant "green" custom homes nourish providers of green technologies with huge infusions of money. As such a provider, I am not immune to this logic. My personal standard for a job of this sort is that if the owners are willing to commit themselves and other users to significant lifestyle accommodation for the Earth and risk pioneering new designs, then I'll do it. Redefining comfort and convenience standards can easily halve the project's environmental impact. Furthermore, it promotes Earth awareness and lifestyle accommodation to an influential class of people. Still, these goals can be accomplished even better in fewer square feet. One of the most powerful lifestyle statements I've witnessed is a very successful attorney and his wife living happily for several years in a tasteful 120 ft^2 *(12 m^2)* cottage on a multimillion dollar lot in Santa Barbara. That's the size of the master bedroom closet of the house you'd expect these people to live in!

Error: Pump Zeal

Many people reflexively reach for a pump when they want to move greywater around. There is no denying that in the initial demo, it is very cool to see greywater squirting out where you want thanks to a pump. But the pump is the first step on a slippery slope toward the Overly Complex System error. It won't work very long without filtration, and filtration is a big hassle. Your house will still be generating greywater in 30 years, but pumps are not a longterm solution. Odds are that if your greywater system uses a pump, it won't still be in use in just five years. If the pump hasn't been killed by hair wrapped around the rotor, or by you when the float switch hung up at 3:00 am and stayed on noisily for the third night in a row, that's about the time it will die of old age. Unless you're really enthusiastic about cleaning fetid filters after doing it hundreds of times (often during parties, when use peaks), you probably won't buy a replacement.

Even if everything works perfectly, the pump sucks a lot of electricity; it is typically the second-most to fifth-most energy-using appliance in the home. So for all your effort, you've just substituted electricity waste for water waste.

Preferred practices: Use a system that doesn't need a pump (see the System Selection Chart).

Exceptions: If all your potential irrigation area is higher than your greywater sources, you have to either forget about reuse or get a pump. In this case, bite the bullet and get a good one that doesn't require filtration (Appendix E). In institutional contexts with more than 300 gpd of greywater generation and a similar irrigation need, the water can easily be worth more than the fancy filtration and pumping hardware needed to manage the large volume, even if it all has to be replaced every five years. At present, volume greywater reuse hardware and support are weakly developed, but this is a legitimate application for a pump.

Error: Storage of Greywater

The word "storage" should immediately sound an alarm, as should any residential system that includes a tank bigger than 55 gal *(200 L)*. Storage rapidly turns greywater into blackwater. If you doubt this, fill a bucket with greywater and observe as it progressively darkens and reeks. Bacteria (at least indicator bacteria) multiply to blackwater levels as well. In Mexico the *trampa de grasa* (grease trap) often included in greywater systems is a popular way to commit this error. Omitting, bypassing, or downsizing the *trampa* would be much better.

Preferred practices: Eliminate pooled greywater anywhere it occurs—send it straight to the soil. The fewer anaerobic corners and pockets the better. My latest designs drain *completely*. The collection and distribution plumbing, and surge tanks (if any), slope at least 2% across their bottom surfaces.

Exceptions: Temporary "storage" in surge tanks, which moderate peak flows, is okay.

Greywater intended for manual distribution can be stored for the day to allow for its distribution all in one session. Tanks for this purpose should drain completely (not leave a bit of fetid greywater at the bottom to inoculate the next batch) and *not be too big* as this invites misuse in the form of letting the water sit too long.

Greywater can be filtered by settling in a septic tank, but then it must be disposed of like blackwater. In this case, the longer it sits in the tank the better (lower suspended solids and BOD; see Settling (Septic) Tank, Appendix E).

Highly treated greywater (for example, from a Constructed Wetland) can be stored for up to a month before it goes septic, depending on the BOD and temperature.

Really cold greywater takes longer than 24 hours to stink, though I'd need a really persuasive reason to rely on this factor in system design.

Beautifully made but ill-conceived greywater storage tank. *This 250 gal (1 m³) tank is the final of a four-chamber system that converts greywater to blackwater over the course of a week. The blackwater smells wicked, and generates hydrogen sulfide, which has eaten through the steel lid.*

Error: Disregard for Real Public Health Concerns and/or Excessive Paranoia about Imagined Health Concerns

Some people advocate irrigating lettuce and carrots with untreated greywater. Others fret about distributing greywater under 9" *(23 cm)* of soil without disinfection. Some worry about eating fruit that contains molecules from biodegraded dish soap, forgetting that they imbibe traces of dish soap directly with every glass of water and plate of food.

Preferred practices: Each user has to find his/her comfort point on the paranoia vs recklessness continuum, and each community has to determine the limits of recklessness it will tolerate. Greywater reuse poses only a very mild health threat in industrialized countries. Proper handling can eliminate the small health threat (see Health Considerations, p. 18).

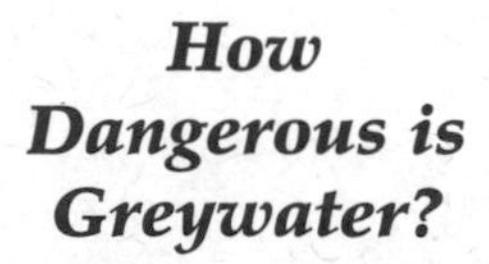
How Dangerous is Greywater?

There are 8 million greywater systems in the US, with 21,000,000 users. In the past sixty years, there have been over a billion user-years of exposure.

There have been zero documented cases of greywater-transmitted illness.

—graywater.org

Error: Treatment Before Irrigation

Pretreatment is sometimes presented as an essential element in a greywater system. In fact it may be more pointless than treating your wastewater before sending it down the sewer. Plants and soil are fine with funky, chunky water. It is pipes and people who may have a hard time with it. **Pretreatment is only necessary to overcome limitations of the distribution plumbing.** With proper design, even straight kitchen sink water can be reliably and safely distributed with no pretreatment whatsoever (see photo on p. 80).

Preferred practices: For simple residential systems:

1. Design the distribution system so it can handle funky water, particularly high levels of solids.
2. Design the distribution system so human (or animal) contact is unlikely to occur before the water has passed slowly through healthy topsoil.

The Branched Drain and other systems fulfill these requirements.

Exceptions: For several hundred gallons a day of greywater with a similar irrigation need, it can be economical to pretreat greywater so it can be distributed by more efficient irrigation hardware.

Error: Greywater Staying on the Surface

This is the opposite of the error above: applying greywater to hard surfaces, to waterways, or in such quantity that it runs rapidly over the surface for long distances without soaking in.

Preferred practices: Contain and cover greywater where it can soak cleanly into the soil.

Error: Surface Greywatering of Lawns

This issue puts me in an awkward position. Do I denounce thousands of greywater users who see no harm in what they are doing (and in most cases are causing no harm), or condone a marginal activity?

Applying greywater to the surface of a lawn short-circuits the all-important purification step. If the lawn receives traffic, it invites direct contact with untreated greywater. The likelihood of transmitting pathogens is small, but it exists.

Greywatering a lawn is also a pain in the rear (see A Note on Lawns, p. 15). The system almost universally used is a hose from the washing machine or house plumbing that is moved all around the lawn. The effort to greywater a large fruit tree is much less, as the target is one big root system instead of thousands of tiny ones. Perforated pipe under a lawn has efficiency in the single digits, and leaves some areas dry.

Preferred practices: It is less risky to use greywater on a lawn that doesn't receive traffic—

Soapy creek from a legacy system *at a hot springs resort constructed in pristine wilderness around the turn of the century.*

A laundry pipe runs out the formal drive of an estate *in the foothills above Los Angeles, and discharges into the public gutter, in which greywater flows untreated for blocks before running into a gulch.*

but then it shouldn't even be a lawn! Lawns are such resource hogs that their only justification is being better to play on than, say, a stone and cactus garden. Solution? Replace the ornamental lawn with something else entirely, and irrigate that with greywater (if it still needs irrigation).

Exceptions: Subsurface drip irrigates lawns safely and well. Greywater generation/irrigation need of 100–300 gpd is the break-even point where such a system, at $2,000 or more, could start to make sense (see Automated Sand Filtration to Subsurface Emitters and Septic Tank to Subsurface Drip, Chapter 8).

Error: Irrigating Vegetables

The reason not to use greywater on vegetables is concern about transmitting disease. Some people greywater veggies anyway. Why? See Exceptions, below.

Preferred practices: If your goal is to get rid of your greywater responsibly and not to irrigate, there is no reason to put greywater on food crops. If your goals are both to grow food and lower overall water consumption, chances are you'll have more irrigation demand than greywater. In this case, use greywater first on ornamentals, then on fruit trees, and use the freshwater you saved to water your veggies.

Exceptions: In the past I have categorically recommended against greywatering vegetable gardens, but I've decided to write more about the boundary between responsible resource reuse and reckless public health threat in this area, because I understand the attraction. My vast drip irrigation system does not adapt well to veggies. The hardware is not great, coverage is uneven, water goes on top of mulch instead of under, and even if I build the beds to match the micro-sprinkler spray pattern, wind blows the water off course. Lines with inline emitters are super expensive, and clog rapidly even from filtered potable water. Additionally, I can't just hook up to one of the other irrigation zones because veggies need about 10 minutes/day, instead of, say, 1½ hours three times a week. During weeding, seeding, transplanting, growth, and harvest, irrigation need varies tremendously. Veggies also are much more sensitive than fruit trees to daily (even hourly) weather changes. In short, home-grown veggies are a poor fit to automated irrigation. The efficiency is likely less than 30%, whereas I'm used to 80% for my fruit trees. Hand watering is the obvious alternative, and if I'm hand watering anyway, why not use hoses or buckets of free greywater, instead of freshwater costing me nearly a penny a gallon? Hand watering with greywater can yield 80% efficiency, and I'm much more tuned into the plants.

Veggies irrigated subsurface with greywater in a greenhouse*—a probable exception to the no-greywater-on-vegetables principle.*

The greywater systems I recommend include several layers of protection, each capable of preventing the spread of infectious microorganisms on its own. But when irrigating veggies with greywater, the tenuous protection is only from:

1. Not happening to have anything nasty in the water
2. Only getting nasties back that came from you in the first place
3. Not splashing greywater on the edible portions
4. Washing the veggies
5. Cooking the veggies
6. Not getting sick even if you eat something nasty, because your immune system is robust

Additionally, greywatering vegetables is typically manual, which inevitably results in direct contact with greywater. If you engage in this reckless practice, pay attention to what you're doing, because your margin of safety is slim. If anyone in the house has an infectious disease, protections #1 and #2 above are not operative and you should stop using greywater on veggies. For protection #3, exercise care in applying greywater, and give crops that are splashable and/or eaten raw (e.g., carrots, salad greens) a wide berth, even more so near harvest time. Try not to splash, and wash your hands afterward. For #4, always wash greywatered raw veggies with soap, iodine, or equivalent. (I'm reluctant even to mention #6, but it seems to work in much of the non-industrialized world.)

Error: Irrigating Plants That Can't Take It

Certain plants cannot take greywater, and certain plants can't take too much of any kind of water. Acid-loving plants tend to have a hard time with greywater. Certain plants native to dry areas have a hard time with any dry-season irrigation. Also, some cultivated plants have problems with soggy conditions.

Preferred practices: Avoid irrigating plants that don't want it and design your system so it doesn't create sogginess for plants that can't handle it.

Error: Garden Hose Directly from the Washer

This works until any of these problems occur:

1. The hose kinks, and, with the flow stopped, the washer pump burns out.
2. The pump, after months of working too hard to push water through a long, skinny hose, burns out.
3. Scalding hot water boils your prized plant.
4. The machine tries to refill itself, and the hose (with its end lower than the washer) siphons freshwater out until someone shuts off the washer.
5. The pump doesn't get all of the water out, so clothes stay soggy.
6. You lift the hose to move it, and weeks-old, fetid greywater rushes backward into the machine and onto your clothes.

Preferred practices: The Laundry to Landscape system solves all of these problems, except #3. The Laundry Drum solves them all (see Chapter 7: Simple, Easy Greywater Systems).

Exceptions: As with many greywater errors, a certain percentage of "erroneous" installations operate for a long time without any of these problems. People may get away with this "error" because of a favorable confluence of washing machine internal plumbing and site geometry. If your installation has worked for years, why rock the boat? On the other hand, if you're installing a new system or replacing your old one, why not choose a design that is sure to work?

Dry season greywater application is suspected of toppling these two giant oak trees *in the most dramatic case I've ever seen of this problem.* ***(****Dry season irrigation promotes root disease in these trees.) Mysteriously, oaks across the street seem unaffected by their greywatering.*

Error: Perforated Pipe or Other Distribution System Where You Can't Tell or Control Where the Water Is Going

There are two problems with distributing greywater through perforated pipe:

1. It will clog—Perforated pipes don't clog in septic tank leachfields because a septic tank effectively removes suspended solids. Suspended solids are plentiful in greywater, even after the crude filtration sometimes attempted in home-brew systems. If solids don't eventually clog the pipe, root infiltration will. If the pipe is big enough, and the greywater clean enough, it may take so long to clog that the durability is acceptable, but this is rare. A seldom used bathroom sink draining into 10' of 4" *(3 m of 10 cm)* perforated pipe in gravelly soil might last several years before failing. Laundry water, however, will quickly clog almost any size perforated pipe.

2. It is unmanageable for irrigation—To actualize water savings, you have to coordinate with freshwater irrigation so all the plants get enough but not too much. This is all but impossible to do with perforated pipe, which invariably waters one small area too much and the rest hardly at all, in a pattern that is constantly changing and, if the pipe is buried, invisible—except in the form of distressed plants.

You may be able to fine-tune the pipe slope, hole spacing, and size so each hole spits out the same amount of water along the entire length. Attempted by numerous greywaterers before you, this perfection is very fragile. If you alter the slope 1% or the flow a few percent, or if lint blocks some holes, distribution evenness is shot.

A series of garden beds with water flowing in gravel underneath is another example. Will the first bed get 90% of the water because the plants suck it up before it can move on, or will the last bed get the most because water rushes down to it? Who knows?

Preferred practices: If you must have perforated pipe, *add it as an extension to your leachfield and filter the greywater through the septic system first so the pipe doesn't clog.*

For separate disposal of residential greywater, use mulch-filled basins supplied by a Drain Out Back or a Branched Drain, with pipes a few inches above mulch or in good-sized underground chambers.

For reuse, plumb in such a way that you know with certainty where the water is going and can adjust your supplemental irrigation accordingly. This typically means a parallel system, or one with only one outlet. Examples are a Drain Out Back, Movable Drain Out Back, Branched Drain, well-made and regularly serviced distribution boxes, bucketing, etc. Crude but effective parallel splitting of greywater flow can be achieved by not combining fixtures, keeping flows separate in the first place. This is difficult to manage if fixture use is highly variable and/or unknown, but works in some applications (see Lump or Split the Greywater Flow, Chapter 3).

For large flows, high-level treatment and underground drip irrigation are preferred (p. 71).

Exceptions: Pressurized effluent with suspended solids removed, for example, by septic tank or sand filtration, *can* be distributed evenly through perforated pipe. For even irrigation, pressurized perforated 1" pipes with 1⁄16" holes on 1" centers up to 50' long have been used (see photo on p. 72).[m]

As with many of these cautions, some folks seem inexplicably to get them to work well; see photo on the right. I've heard reliable accounts of a few gravel pit or trench systems that have worked for ten years or more with shower water only. Perhaps there is something magical about these installations, or shower greywater in general?

Laundry water is normally the worst. However, I have seen an old-fashioned clay drain tile line that handled laundry greywater for 30 years in clay soil without clogging. Although it was not failing, it was replaced with modern perforated pipe as part of a remodel. The replacement clogged completely in less than a year and was abandoned—go figure.

An exception: "Twisted tees" system in Santa Fe, NM has been successfully tuned so that the same amount of laundry water comes out each outlet, each of which branches off at progressively lower angles toward the end.

Error: Combined Wastewater Designs Used for Greywater

Most plumbers have more experience with combined wastewater (greywater + blackwater) than with dedicated greywater plumbing, and sometimes misapply principles of one to the other. Here are some common pitfalls:

Pipe Too Big

Combined wastewater in residences generally requires 3" or 4" pipe. This is *too big* for greywater. If you just ask your plumber to run greywater and blackwater separately until outside the house, you'll likely end up with a 3" or 4" greywater stub out. How can a pipe be too big? First, it is more awkward, more expensive, and burns up valuable elevation faster. Second, crud in small flows doesn't flow down it as well (see Figure 4.7).

Preferred practices: Use 1½" or 2" pipe.

Exceptions: Huge multifamily or institutional flows.

Pipes Too Low

Combined wastewater in residences generally is on its way to sewer or septic lines deep underground. This is *too low* for greywater.

Preferred practices: Plumb pipes as high as possible—see Squander No Fall, Chapter 4.

---

[m]*Metric: 25 mm pipes with 1.5 mm holes on 25 mm centers up to 15 m long.*

Error: Freshwater Designs and Hardware Used for Greywater

Although it may seem natural to expect greywater to follow the same laws of physics as water, it doesn't. Here are some common pitfalls:

U-Shaped Pipe

Greywater should pass through a U-shaped pipe section, seeking its level just like freshwater, right? Wrong! Crud in greywater settles at the bottom of U's, clogging them.

Preferred practices: Continuous slope of 2% in rigid lines. This prevents airlocks as well, which plague all low-pressure inverted U-pipes, freshwater and greywater alike.

Exceptions: Flexible lines won't develop this problem if moved once in a while.

Pressurized greywater lines blast the crud through U's.

Traps are short U's at the bottom of vertical drops that (usually) blast crud through.

Cheap Electric Valves

Greywater distribution can be controlled elegantly and automatically using $15 drip irrigation electric valves, right? Wrong! Crud in greywater prevents the valves from closing, greywater quickly corrodes them, and greywater rarely has enough pressure to make this type of valve (powered by water pressure) work right, even on the first day.

Preferred practices: Forget automated valves, use one of the handful that work with greywater (see Valves, Chapter 4), or buy super-expensive electric sewage valves.

Filters

This nifty sand / gravel / carbon / reverse-osmosis / fill-in-the-blank filter will filter my greywater just like it does my freshwater so I can… Wrong! Crud in greywater clogs that filter in the blink of an eye.

Preferred practices/Exceptions: Don't filter greywater, or see Appendix E for proven greywater filtration options.

Soaker Hose

I'll just run my greywater through this soaker hose… Wrong! Soaker hoses are such a poor technology that they typically have 30% variation in flow when brand new and used with freshwater, and soon clog with all but the cleanest freshwater, let alone greywater.

Corrugated Flexible French Drain Tubing

Corrugated flexible drainpipe has small, rapidly clogged outlet holes and collects festering crud in all its ups and downs, as well as the corrugations, which are likely to elevate bacteria and smell levels.

Error: Inexpensive Greywater to Drip Irrigation

This subset of the error above deserves special mention. Backyard tinkerers tend to converge independently on the idea of distributing greywater through drip irrigation hardware. How else to achieve 80% irrigation efficiency, short of manually bucketing water plant by plant? The only problem is that it doesn't work. Of the hundreds of such systems built during the California drought of the 1990s, every one I know of was abandoned within a few years.

The most common configuration was a surge tank with an inlet filter and a float-actuated sump pump to pressurize the lines (see Drum with Effluent Pump, Chapter 8). It works great for the first few weeks, then the filter clogs…

At first, the drains are just slow, but then not draining at all. Finally, there's no denying it's time to clean that filter. When you remove it for cleaning (wearing rubber gloves, of course), you realize too late that the upstream pipes are full to the brim. At this moment, a deluge of chunky, days-old, backed-up greywater pours into the surge tank. Fortunately, you are quick on your feet and only lightly spattered. Not more than a third of the solids caught in the filter have spilled into the surge tank. As you struggle to get your gloves off to wipe the flecks from around your lips and eyes, the sump pump cycles on. Horrified, you stand there paralyzed for a moment as $300 of dripline is pressurized with last week's split pea soup. By the time you wiggle under the crawl space and pull the pump's plug, the dripline is history and the pump is already stopped by hair wound around its rotor anyway.

Preferred practices: Do something simpler and more robust with less efficiency, choose a more labor-intensive, less sanitary system like bucketing, or bite the bullet and pay $2,000+ for a greywater-to-drip system that works (see Automated Sand Filtration to Subsurface Emitters, Septic Tank to Subsurface Drip, Chapter 8).

Exceptions: None.

Error: Low-Volume Reuse Systems for Toilet Flushing

Flushing with untreated greywater fouls the tank and produces fetid smells (see Error: Storage of Greywater). Treatment is complex and expensive, as is automation. The cost for one system on the web for greywater-to-toilet-flushing is $10,000. Even at the punitive water rate of $0.01 a gallon, that's five years of 325 flushes a day to recoup your investment, not counting lost interest, electricity, or system maintenance. The $650 (2007) Homestead Utilities system,[s21] the cheapest I've heard of, would take 23 flushes a day. In a restaurant, it could earn its keep.

Economic infeasibility can indicate ecological infeasibility; the Earth could be much better off if you just wasted the water than if you wasted all the plumbing, pumps, tanks, filters, and electricity needed to make this sort of system work.

Preferred practices: Put in an ultra-low-flush toilet (or a waterless composting toilet; see oasisdesign.net/compostingtoilets/book). Second, "If it's yellow let it mellow, if it's brown flush it down." Third, toilets can be flushed with greywater by simply bucketing it directly from the bathtub/shower into the toilet bowl (*not* the tank). An added plus of reusing bathtub water in this way is that since the flush volume is under intelligent control, you can use just the right amount of water needed to get the job done. Also, in cold climates, you get a primitive but highly effective sort of greywater heat recovery as the bathwater heat escapes into the home. Finally, there's the option of installing a "lid" that flushes the bowl with cascaded handwash water (see photo on p. 137).[s7]

If none of these options appeal, you're better off forgetting about flushing with greywater.

Exceptions: If you have highly treated water already, say from a Constructed Wetland, and don't know what to do with it, it may be worth supplying it to the toilet. An advantage is that toilets need flushing every day, whereas irrigation need is usually seasonal.

Clearwater such as air-conditioner drip, reverse-osmosis water purifier reject water, and fixture warm-up water is a natural for flushing toilets. It needs no treatment and can be stored indefinitely. Steven Coles of Phoenix, AZ suggests that supplying the toilet from the reservoir of an evaporative cooler also keeps the mineral concentration in the cooler water from rising.

Multifamily, institutional, or any other high-use installations can benefit from flushing toilets with highly treated greywater, especially when incorporated in the original design of the building.

Error: Use of Government Agencies, Trade Organizations, Engineering Firms, or Salespeople for Greywater Design

There is very little overlap between the set of practical greywater systems and the set of legal greywater systems (outside of Arizona, New Mexico, and Texas). This seriously hampers the government's ability to give out useful information, if they had it. Because greywater reuse is a new, marginal, rapidly evolving field, it is hard for ponderous bureaucracies to keep up. California greywater law and the pamphlet that explains it are especially misleading for hardware guidance (see error below).

Also, few practical greywater systems can be profitably installed professionally. It is likely that an established trade organization, engineer, or plumbing company would set you up with an unproven, overkill system adapted from some better-known treatment technology.

Appropriately designed greywater systems are not a very sales-friendly product: way too site specific, variable, and inexpensive. The Internet features numerous generic, expensive, prepackaged greywater systems with fantastic claims, which seem to benefit the vendor more than the user.

Preferred practices: As far as I know, our books and website are the most complete and up-to-date references for home greywater systems. The designs in our books promise less than in other sources…because we stick closer to reality.

Exceptions: For new construction, sometimes an over-engineered greywater system is less over-engineered than conventional treatment. Also, a few prepackaged greywater systems are actually good[s5] (please let us know if you make or have used a good one).

Error: California Greywater Law Used as Model

California's greywater law is an important step and certainly as well done as was politically possible. Too bad it's a step not quite in the right direction, as it is being widely emulated. Some of the hardware recommendations are questionable. The Mini-Leachfield system, for example, is described in great detail as if it were a proven technology, but has been installed in no cases I know of. I can't think of any application for which I would recommend it, either.

Unrealistic laws have poor participation rates. Santa Barbara, for example, issued approximately ten permits for greywater systems in 1989–2009. Yet there is evidence that during this time, which included several years of severe drought, over 50,000 Santa Barbarans used greywater! There are so many obviously overkill requirements that the entire law, including the sensible provisions, is dismissed as a source of design guidance.

Preferred practices: If you're a homeowner, don't follow the CPC/UPC-style law unless you can get a favorable interpretation from your inspector or have no choice.

If you're an inspector, sections of the law grant you nearly total discretion to approve whatever you want. Please exercise this discretion to discount the more pointless sections of the law and allow genuinely well-designed and executed systems. (There's more on this law in our *Greywater Policy Center*[4] and it is discussed line by line in our *Builder's Greywater Guide*.)

If you're making policy, don't blindly follow the CPC/UPC's lead when writing your own administrative authorities' greywater regulations—instead, follow Arizona and New Mexico. A reasonable regulatory stance leads to a greater compliance rate and reduces the risk from perpetuation of unregulated systems.

2009 update: The California greywater law is being updated as we go to press. See graywater.org for the latest information.

That's it for greywater basics. Fill out the checklist on p. 121 if you want, or just get your system going...

Chapter 12: Real World Examples

We suggest you scan through these and read at least the ones that are similar to your context in some way…though something can be learned from all of these examples.

So far we've started with general principles and explored their application in different contexts. Now we'll go in the opposite direction: starting with specific contexts and seeing how the principles get applied. Some of the examples below are exotic and extreme, all are real. There's something to learn from each of them. Their titles give the most salient site feature, the location, and the greywater system type.

Example #1: Town—City of Santa Barbara (Branched Drain)

A classic residential Branched Drain system on a small lot in town with a really nice garden.

System: Branched Drain with dipper, to sub-mulch emitters.

Context: Part of a State of California study.[30] (See *Builder's Greywater Guide*, Getting a Permit, for a copy of the permit and site map.)

Goals: To enable yet more extensive gardening without raising the water bill (the original owners were keen organic gardeners), to send the greywater somewhere other than the ocean (a recent spate of beach closures had raised public awareness about this cost of sewer systems), to gather data for a greywater study, and to demonstrate feasibility of Branched Drain system design in an officially monitored and permitted installation.

Under these fruit trees are inverted 5 gal pot outlet chambers *in large, deep basins. Basins are filled completely with wood chips. The system is completely invisible in the landscape, yet easily serviced by pulling mulch back.*

Branched Drain to Mulch Basins *that I installed in Santa Barbara as part of a State of California study.[30] (There's a bigger picture on the front cover.) The study showed that greywater systems save water and do not adversely affect soil quality. This system has received only occasional inspection since installation, and reliably distributes greywater to about a dozen fruit trees.*

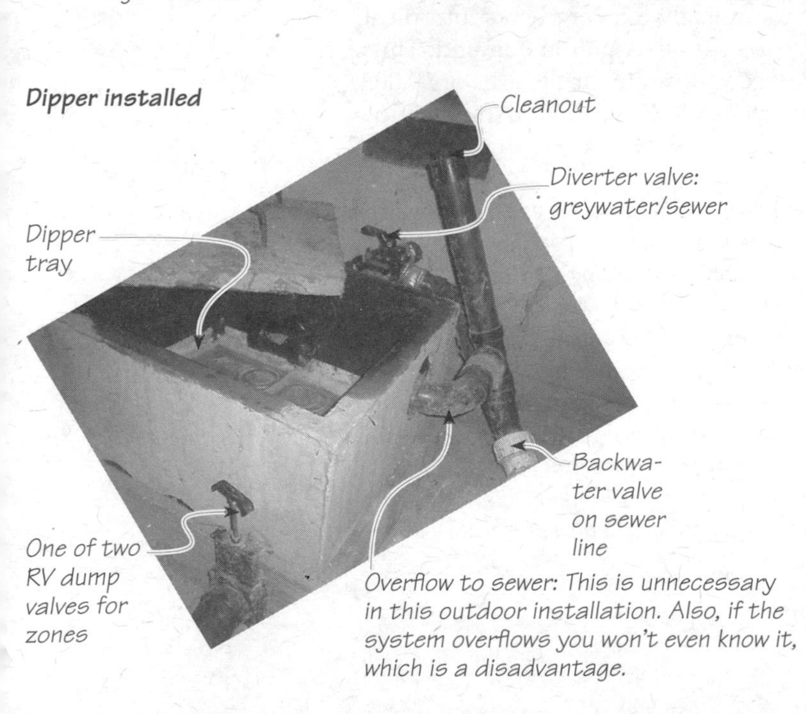

Dipper installed

Greywatered bananas with outlet under clay pot

Design and installation issues: Conserving fall was especially arduous on this site, which has an average slope of 2%. Because greywater pipes need a 2% slope, this sounds great, until you realize that when you go at an angle across the slope, the pipes slope less. Maintaining adequate slope was greatly complicated because we threw away perfectly good fall at many points to comply with the 8" *(20 cm)* legal burial depth requirement for the pipes (I don't think the inspector even noticed).

The basin area is way oversized, as per California greywater law.

Costs and benefits: The study reported favorably on the economics of this system—despite the fact that its cost was nearly doubled by the legal requirement to plumb twice as much area as there was water to irrigate with, and nearly doubled again by the requirement that the flow be measured. It cost about $1,000 in parts and $360 for a plumber to help re-plumb the entire underneath of the house to gain the 8" fall needed to run a dipper box with a counter. The other two systems in the study handled similar amounts of water but cost $5,000 each and had an estimated five year life expectancy. The life of the Branched Drain system was estimated at 20 years (1998 numbers). The study also showed that greywater systems saved water and did not adversely affect soil quality.

Example 1 costs from 1998.

The system has received only zone switching and slug control since installation, and reliably distributes greywater to about a dozen fruit trees.

User's view: The house's new owners are maintaining the beautiful garden nicely. They were very concerned about the greywater system before buying the house and are pleased to have had no problems, or hardly anything to do about the system, in the years since. They said: *"We never saw water surface. Not even from the lowest outlet, during a deluge of rain when we turned on all the fixtures in the house for a few hours to test the surge capacity of the system."*

Example #2: Suburbia—Southern California (Branched Drain)

Another classic Branched Drain system on a relatively small, intensively planted lot.

System: Branched Drain to free flow, sub-mulch, and subsoil emitters (see figures on facing page).

Context: Three separate Branched Drain networks irrigating an existing, extensive orchard. Mostly owner-installed one at a time starting in the kitchen, with some hired labor for trenching. Supplemental irrigation by automated drip.

Goals: To enable the orchard to grow without using even more water from a limited supply, and to stave off replacing the septic tank leachfield.

Costs and benefits: Costs are summarized in Table 12.1 (a simple cost/benefit analysis for this system can be found in Table 2.7). With 85% of the wastewater diverted from the septic, it is almost inconceivable that the leachfield will require replacement within the owner's lifetime. This saves a future expenditure of several thousand dollars. The septic tank has been monitored but has not required pumping since this system was installed. The interval between pumpings is probably doubled. One pumping at a cost of $300 has been saved thus far. Irrigation efficiency is high because the greywater goes mostly to evergreen subtropical fruit trees at a rate approximately equal to their average *wet season* irrigation demand. Thus, there is little water applied that is not useful. At an average water rate for the area of $0.003/gal *($0.80/m³)*, the three greywater systems together save about $68 a year, total, for a simple payback time of eight years on water savings alone, or two years considering the septic savings.

Example 2 costs from 1998.

User's view: *"I installed the first Branched Drain system unglued because my experience was that greywater systems were unreliable. I was skeptical that the flow splitters would work for even a day before one side clogged and all the flow went to the other. I tried very hard to clog it at first but it never clogged, not even where the flow was split into 16ths! As the superiority of the kitchen system became evident, I put in the other Branched Drain systems one after the other. If I had done them all at once, it would have been better to join the flows before splitting them. We leased out our house for a year and the system did great without even being looked at. The only difficulty was one separated joint. I'd have glued the first system if I'd known it was going to work. The new systems are glued. The systems are low maintenance. We add wood chips annually and extend basins as fruit trees grow—things I did before there were any greywater systems. We've got a few way-undersized subsoil outlets (5 gal pots) that I've mucked out once a year. These should be replaced with properly sized ones. Overall, I am frankly amazed at the robustness and self-cleaning of this system type."*

TABLE 12.1: COST OF SUBURBAN GREYWATER SYSTEM

Subsystem	Parts	Plumber hrs/$	Laborer hrs/$	Owner labor hrs
Kitchen	$100	1/$50	0	20
Bath	$150	3/$150	10/$80	15
Laundry	$150	1/$50	10/$80	18
Subtotal	**$400**	**5/$250**	**20/$160**	**53**
Total cost	**$810**	**(1998 costs)**		

Four outlets with 1/8 each of bath/shower flow. These subsoil outlet chambers are way undersized. The owner plans to replace them with a continuous row of infiltrators 30' (10 m) long and 14" (35 cm) wide.

FIGURE 12.1: GREYWATER SUPPLY, IRRIGATION NEED, AND DISTRIBUTION SYSTEMS

Laundry system irrigation need:
245 gpw now,
1,335 gpw in 10 yrs
192 gpw greywater generation

Bath/shower system irrigation need:
237 gpw now,
318 gpw in 10 yrs
254 gpw greywater generation

1' (30 cm) graphic contour lines

	Metric
5 gal	19 L
192 gpw	0.73 m³/w
237 gpw	0.90 m³/w
245 gpw	0.92 m³/w
254 gpw	0.96 m³/w
261 gpw	0.99 m³/w
318 gpw	1.2 m³/w
357 gpw	1.4 m³/w
799 gpw	3.0 m³/w
1,335 gpw	5.0 m³/w

Kitchen system irrigation need:
357 gpw now,
799 gpw in 10 yrs
261 gpw greywater generation

Deciduous tree   Evergreen tree

Darker shaded areas have higher irrigation need. Unshaded areas are irrigated occasionally or not at all. Supplemental irrigation is by drip irrigation on a timer.
gpw = gal/wk

Outlet chambers are upside-down 5 gal buckets with access holes drilled in the bottom. The square tile is the lid over the access hole.

Two outlets with 1/4 each of bath/shower flow.

Example #3: No Water—Highland Central Mexico (Various)

A high class system in a non-industrialized world context with acute water shortage.

Systems: Branched Drain to free flow outlets, Multi-Mode Greywater Tank, septic, and Constructed Wetland to Branched Drain.

Context: A large home in an international ecovillage at 7,000' *(2,300 m)* in central Mexico. The village has an eight-month, dust-dry season followed by 4½' *(140 cm)* of rain in four months (!). The only dry-season water supply is from cisterns filled during the monsoon. Freshwater irrigation during the dry season is unfeasible, as cistern storage is prohibitively expensive; enough capacity to provide one person with 5 gal *(20 L)* a day for the dry season costs $2,000—what a laborer makes in a year. Villagers without sufficient cistern capacity travel two hours round-trip by donkey to distant springs to carry back 10 gal *(40 L)* of water, collected drips at a time. As in most of Mexico, Huehuetortuga's greywater is generally dumped through Drains Out Back, or simply on the ground wherever it is used, with poor sanitation and inefficient or zero reuse.

Huehuetortuga house and mountains.

This site has a dry toilet for the owner.[31] It also has a large septic tank, and a tiny Constructed Wetland which treats the water from two low flow toilets (see photo in Constructed Wetlands, Chapter 8). Greywater systems include a Branched Drain network for two kitchen sinks and a new hybrid design I call a Multi-Mode Greywater Tank. It is working well, but trials have been limited. The Multi-Mode system can be a good choice where water for gardens is precious, as long as the users are not prudish about hygiene and have energy to bucket greywater.

Goals: To pioneer/demonstrate new systems for doing more with less water in this area, enable tastier varieties of fruit trees to be grown, and create green relief in an environment where there is no water income of any kind for at least half the year.

Example 3 costs from 1998.

Design and installation issues: It was not known what sort of landscape this house would have or how much the guest rooms would be used, so the systems were designed for maximum flexibility. The kitchen system—the Branched Drain network—is the only system that commits the water to one set of outlets. The trees it serves are the main irrigation commitment. The other systems can all be adjusted and adapted as needed.

The Multi-Mode system has four use modes, depending on the irrigation efficiency desired:

- **Bucketing**—80% efficiency, considerable effort. Pairs of buckets are filled from four tanks spaced around the landscape and manually doled out to plants. It takes about one 5 gal bucket per square yard of irrigated area to apply 1" of water; 1" per week is a typical irrigation level.[m] A benefit of bucketing is that whoever empties the tanks is acutely aware of how much water is used indoors, which can improve indoor conservation. On average, people who carry water *to* their homes use 10% as much as people with piped water. Carrying water *out* should have a similar effect. Pipes can't distribute wastewater to plants nearly as well as they can supply potable water to the home, so if bucketing is going one way or the other, from house to plants is preferable.
- **Manual hose**—65% efficiency, medium effort. Water is released from the tanks once a day through a large-diameter hose with a valve at the end. The hose is moved by hand and shut off between watering areas.
- **Semi-automatic hose**—50% or less efficiency, low effort. The drain of each tank is left open, so it functions as a surge tank rather than as a storage tank. Water runs immediately out the hose to whatever plant is at the end of the line. The line is moved periodically among plants.
- **Automatic hose**—Nearly 0% efficiency, no effort. For the wet season, the hose is routed to a mulch-filled infiltration basin and left there (the soil perks very rapidly, so purification capacity is not an issue).

Multi-Mode Greywater Tank *with enough fall to fill a bucket from a valve at the bottom. A quarter-cylinder shape, this tank fits nicely in an inside corner of the house. It holds 50 gal (190 L) and is made of ferrocement faced with stone. The floor slopes strongly toward the outlet, so it drains completely of water and is self-cleaning of solids.*

The greywater tanks are conveniently spaced around the house, near thirsty plants. The tanks are sized to accommodate one day's high average greywater flow. Tank shape depends on the amount of fall available. Where there is enough fall to fit a bucket under the drain, the tank can be any shape (a plastic drum works), and buckets are filled by placing them under the drain line (see photo at right).

---

[m]***Metric:*** *One 20 L bucket per 1 m² is about 2.5 cm of water.*

Where extra fall was not available, the tank was built with a special "canoe" shape, which can be emptied quickly and completely by bucket (photo at right).

We were able to make a low area next to the tank for people to stand and scoop buckets without having to bend over so far.

The Branched Drain requires only annual flushing near the end of the wet season. The use of the other systems is quite labor intensive in bucketing and manual hose modes, but labor is inexpensive here and many people need work. An excellent wage for a day's skilled work in this area is $8. It was a key design assumption that a gardener/maintenance person would be present to manage the system. Managing all the irrigation takes about a couple hours per week in the dry season—34 buckets a week, on average. (The gardener empties them all by yogurt container onto the plants to ensure the water goes exactly where he wants it to.)

FIGURE 12.2: FERROCEMENT BUCKET-SCOOP ("CANOE") TANK

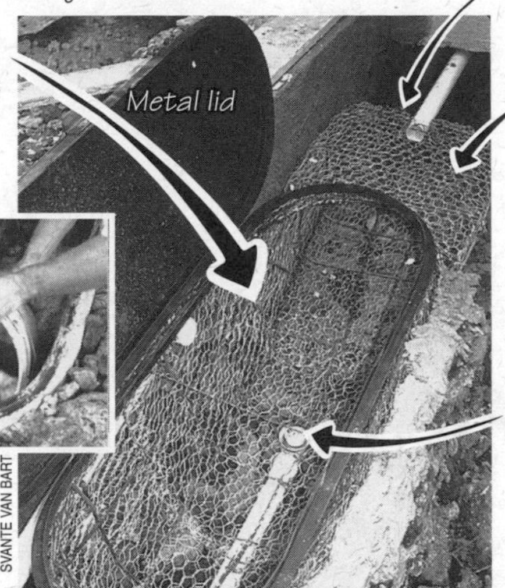

Overflow
Locate at highest point.
Overflow joins a drain,
or just dumps to a nearby fruit tree.
Inlet
Locate as high as possible.
Shape and texture
Tank is ferrocement with smooth finish. The entire tank slopes 1/4" per foot (min) to drain. Bottom and end curves match lower third of a 5 gal bucket to within 1/16". This enables all water and crud to be scooped out very quickly and efficiently. Lid opening should be 20" x 55", with round ends to accommodate a natural bucket scoop. Depth can be varied; 20" is a good figure. Usable volume of tank shown is 30 gal.
Metal lid
Tank extension
To attain daily greywater volume, shape and location of extension are unimportant as long as it drains to low point.
Drain
Locate at lowest point, in middle of bucket scoop. Plug with recessed rubber stopper in bathtub drain with metal cross removed.
SVANTE VAN BART

Costs and benefits: All the wastewater reuse systems together cost about $5,000. This includes the septic system and wetland for treatment and reuse of blackwater. Most of the cost was for design work at industrialized world rates. In addition, the design and construction labor was increased two to four times because most of the project consisted of prototype designs invented for this site, or designs that were new to the region.

WARNING!
Unattended buckets with even a few inches of water are a drowning hazard for little kids.

There are a number of very different ways to evaluate the economics. The least favorable is to assign zero value for treatment and compare the entire cost of pioneering all the systems against the cost of replacing the actual greywater reused so far with the cheapest short-term water source (trucked-in water). Due to low occupancy of the house this past dry season, only about 1,000 5 gal buckets *(20 m^3)* of greywater were generated and reused. (The low demand inside freed an equal amount of fresh cistern water to be used directly in the landscape.) Despite the staggering environmental costs, water trucks can be hired to haul water up from a rapidly dwindling aquifer far down in the valley below for a spectacularly low sum: 1.5¢/gal *($28/7 m^3)*. By this measure, $80 worth of water was saved at a cost of several thousand dollars.

From a more favorable view, the very high reuse efficiency enables the maximum water consumption of the house plus yard to be nearly what it would be with double the cistern capacity. Thus, it functionally substitutes for about $4,000 worth of storage and $2,000 worth of septic system, with much lower environmental impact—90% less cement, for example. Additionally, this system provides extremely effective treatment for blackwater and greywater in a area where water quality is important (all above in 1998 dollars).

User's view: *"My system is a delight. Living in an area that has an eight month long dry season, reusing water is clearly the right thing to do. The four fruit trees that are watered directly from the kitchen sink via a branching greywater system are especially happy with this energy efficient (works on gravity), simple to use, very low maintenance arrangement."*

(More photos and figure of this system on next page.)

FIGURE 12.3: HUEHUETORTUGA LANDSCAPE PLAN
SHOWING IRRIGATION NEED AND FOUR OF SIX GREYWATER SYSTEMS

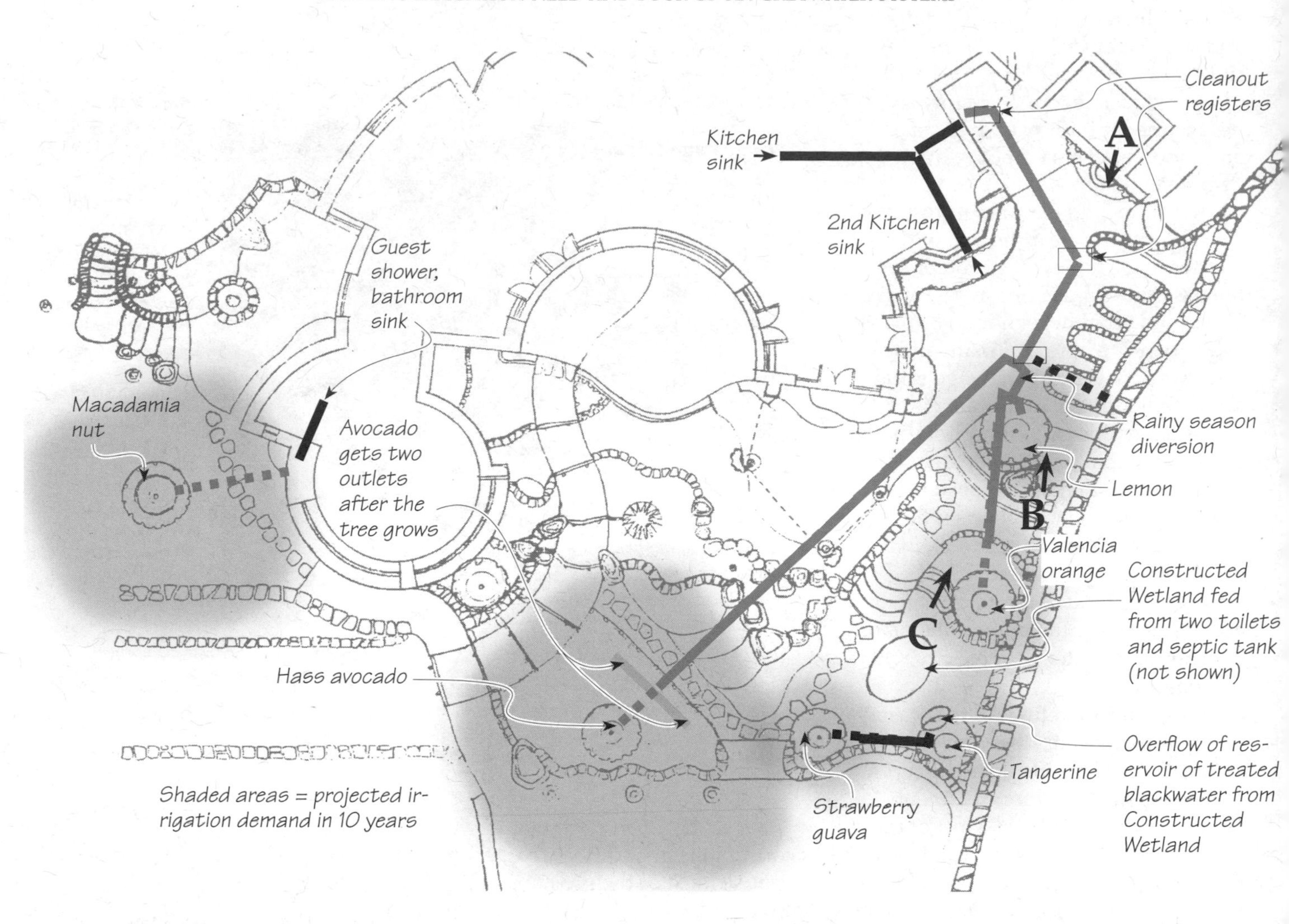

Pipe locations in Huehuetortuga landscape.
Positions and orientations of photos A–C are noted on plan above. Path of pipes is shown on photos with dashed lines. It is best to take pictures of pipes before and after they are covered.

Example #4: Too Wet—Oregon (Mulch-Covered Bog)

A good example of a simple system for responsible treatment of greywater in a wet environment.

System: Drain to two long, sloping, open-ended mulch basin infiltration beds. These are designed to dispose of the water responsibly in the wet season. The water is reused in the dry season.

FIGURE 12.4: EVAPOTRANSPIRATION / INFILTRATION MULCH BASIN IN OREGON

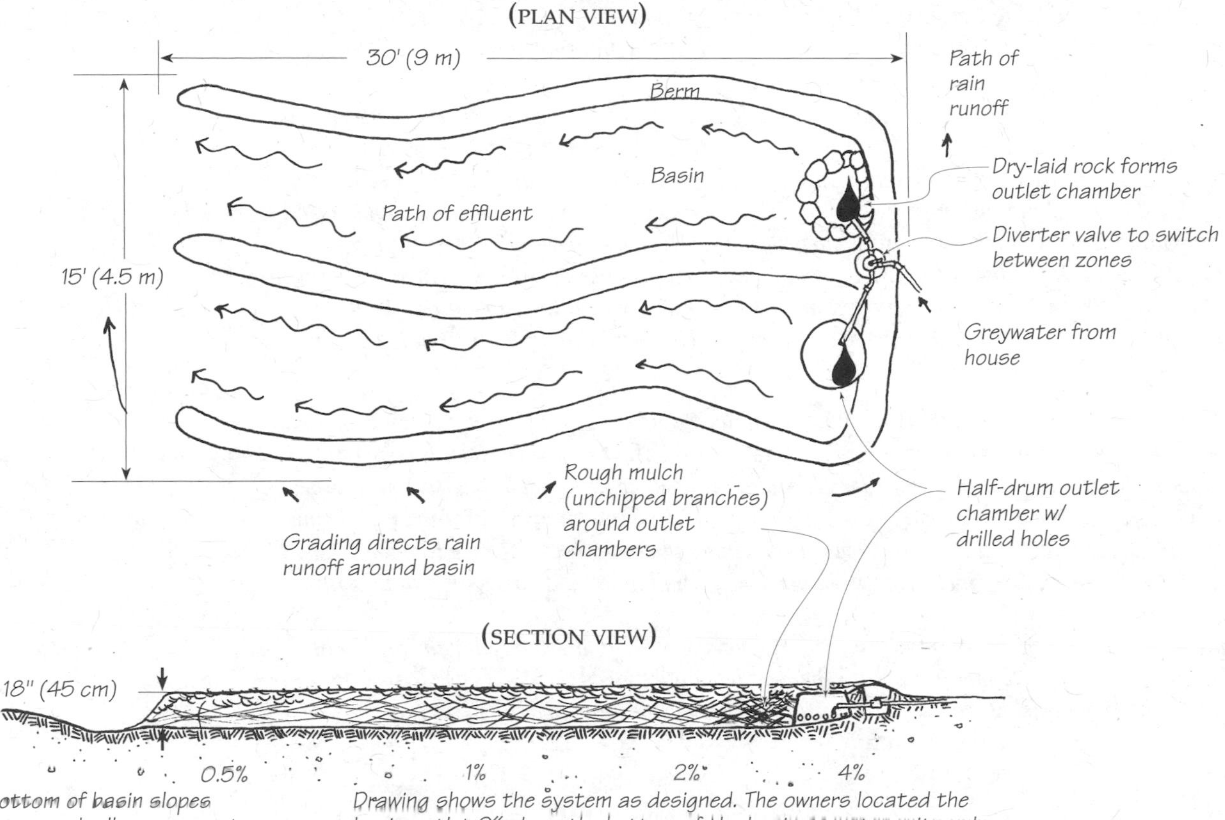

Evapotranspiration/infiltration mulch basin *in Oregon.*

Context: The location is desert in the summer, rainforest in the winter. The owners had already converted to composting toilets, so the only wastewater going in the septic was greywater. However, the septic effluent was somehow leaking into a culvert and thus into a creek nearby.

Goals: To keep the septic tank water out of the creek, eliminate standing or surface runoff of untreated wastewater, and maximize reuse of the nutrients in wastewater.

Design and installation issues, owner's view: The owner had this to say.

"We varied from the original plan in a couple of ways, which I would do differently next time:

1. *We filled the basins completely with bark mulch (not "rough mulch"), and it seems to be compacting down more than I would like. I think that if we had used some rough mulch (unchipped woody sticks) at the bottom, it would have stayed looser longer.*
2. *We put a lot of effort into making the floor of the basin have exactly the right slope. In retrospect, I don't think we needed to have worried about that so much, since I doubt that the water is following the exact path we planned.*
3. *We made the outlet chambers a bit smaller than originally planned, since we were using materials on hand: large planters (approximately 18–22" in diameter) with the bottoms cut off and holes drilled into the sides. We covered these with wooden caps. If I were doing it over again, I would use larger bins for the outlet chambers and put bigger holes in the sides and pack larger, coarser mulch around them. Drainage out of those is a bit slow, and I suspect the water sometimes bubbles out the top, under the wooden caps. Originally, we left the wooden caps exposed on the surface, which resulted in the area smelling slightly anaerobic. The smell vanished when we covered the caps with a thin layer of mulch. I have never seen water surface above this thin mulch layer, although the area is usually very damp.*

We added the walkway and bench so that the mulch basins would look more like a garden area than a water treatment area. We put the most water-loving plants at the inlet end of the basins, since the outlet end often stays dry during the dry months. We switch beds once a week using a diverter valve. I am glad we chose to put in the diverter valve instead of having to swing an open unglued pipe from basin to basin. It makes switching beds so much easier and more pleasant, and it's nice to have the inlets all hidden away. We have the diverter valve housed in a small handmade valve box with a lid, which looks like a bench."

"So far, I haven't seen any water standing in the mulch beds or being discharged from the basin drain, except for what I presume to be rain during the heaviest downpours.... We love our system. Our land is cleaner, and the system is discrete and beautiful. It's great fun on nice days to sit out on our 'observation bench' and watch the bathtub drain."

Example #5: Big Family in New House (Automated Subsurface Emitters)

This example describes a system appropriate for new construction and moderately high flow.

System: Automated sand filter to subsurface emitters, manufactured and installed by ReWater Systems[s5] at the time the house was constructed.

Context: This permitted system is in a neighborhood of large homes with extensive landscapes. Irrigation is required year-round due to low rainfall, semi-desert conditions (10", *25 cm* average annual rainfall). The family has six children and generates lots of greywater.

Goals: Irrigate landscape without excessive water use, dispose of water safely, reduce ocean pollution, save money on water bills.

Design and installation issues: The soil is heavy clay. Tapwater is expensive, salty.

Costs and benefits: This system cost a few thousand dollars for the collection plumbing, and an equal amount for the distribution plumbing, which is being installed piecemeal by the owners. This system qualified for a state Wastewater Recovery Facility rebate, on the strength of ReWater's detailed life cycle analysis. Unfortunately, this program was terminated before the system was completed.

User's views: In my opinion, this site perfectly fits the profile for an automated system. It seems the owners see it that way, too. Wife's eye view: *"I was against it at first. It seemed like a difficult system to set up. I've installed most of the emitters myself. Now that I'm through the learning curve I like it. We have a large lot—a couple acres. We have six children, so we have a lot of water to reuse. Our water bill is not bad at all compared to our neighbors'. I think we're saving a lot of money. And when my 17 year old spends 20 minutes in the shower I don't get nearly as upset as I would otherwise."*

Husband's eye view: *"It is a very sophisticated, well-designed system, but out of reach of most people without government assistance. The system is complex for the average guy. It takes a certain* *amount of maintenance, which is hard as I'm quite busy. We had to replace a pump (a fluke, apparently) and have had to service the filter once in five years. But I love the idea of not wasting the water, and our palm trees are growing faster than anybody's. We have expanded the system to service 25 more fruit trees. The worst thing about the system is the meddling of the water district. We have two expensive backflow valves on the makeup waterline and they need to be re-certified each year at a cost of $75."*

REWATER SYSTEMS, INC.

ReWater surge tank, sand filter.

Example 5 costs from 2005.

REWATER SYSTEMS, INC.

Greywater irrigated landscape, freshly installed.

Example #6: No-Money Ecotourism—Michoacán (Greywater Furrow Irrigation)

This example describes an extremely low cost system relevant to those working or living in the non-industrialized world. This particular system can handle tourist season shock loads 20 times the normal loading rate. You'll find it with the other non-industrialized world information in Appendix D.

Where to Go for More Information

This book's appendices cover specialized greywater topics such as:

- Site assessment
- Measuring elevations and slopes
- Cold climate adaptations
- Systems for the non-industrialized world
- Pumps, filters, and disinfection
- Cascading greywater to other indoor uses
- Sensible regulations
- Units and conversions

You'll find more information on the following specialized topics in our *Builder's Greywater Guide:*[6]

- Installation of greywater systems in new construction
- Conforming with the CPC/UPC
- Improving greywater laws
- Multifamily, commercial, and institutional scale systems
- Greywater science and research

Our website, oasisdesign.net, includes more information on:

- Indoor greywater reuse (for example, for flushing toilets or cooling)
- Greywater policy
- Links
- Water system design
- Greywater for camping
- Plants for greywater systems

There's also a survey where you can provide us with feedback on this book: oasisdesign.net/catalog/satisfaction.htm.

The Future of Greywater Use

In the future, we can look forward to greatly expanded water recycling and options for legal greywater reuse. Underlying necessity has a way of eroding the attitudes and regulations that stifle advances in design and their adoption.[4]

In the US, the reclamation of treated sewage for irrigation is exploding in popularity. Commercial car washes and many industrial processes already purify and reuse their wastewater. In Europe, highly treated sewage is pumped back into the potable water system. These systems require considerably more energy, chemicals, and infrastructure than onsite systems for the reuse of untreated or partially treated greywater.

As global resource constraints tighten, home greywater systems will improve in efficiency, ease of maintenance, and popularity. I predict that such systems—widely illegal only recently—will soon be mandated for new construction in some areas.

Taking responsibility for the small part of the global water cycle that flows through your home yields great satisfaction. I hope this book enables you to benefit fully from the experiences others have had with greywater use, whether you're implementing a proven design or picking up experimentation where others left off. If you experience a notable greywater success—or failure—that you think should appear in the next edition of this book, please drop me a line, or send photos.

Art Ludwig

Oasis Design

oasis@oasisdesign.net

Appendix A: Site Assessment Form

Site ______________________

Date__________

Free copies of this form may be downloaded from oasisdesign.net/design/consult/checklist.htm.

This form serves as a reminder of the common variables to take into account when designing a residential greywater system. Many of these questions may be difficult or impossible to answer—just skip over them for now and they may become clear later. Reading the text of this book will help identify which items are most germane to your context. Also remember that the map is not the terrain...this sheet matters only to the extent it helps things get built well on the ground.

If you are working with other people on the project, this form can serve to orient them, along with a site map and (if they can't visit in person) photos.

Reminder: Careful attention to the context will pay good dividends. Greywater systems are affected by more variables than most systems in natural building, and to a greater degree. A change in one of any number of variables can change the whole design.

Goals

General project goals

What are the guiding philosophies and aesthetic? (E.g., fancy gated subdivision, shack in hippie commune)

What perfection standard* are you aiming for?

Hygiene standard?

Greywater system goals (check all that apply)

- ☐ Irrigate/Save water *(don't forget conservation before reuse)*
- ☐ Dispose of water safely
- ☐ Improve sanitation
- ☐ Reduce pollution
- ☐ Save septic
- ☐ Save money
- ☐ Feel good
- ☐ Demonstration *(it should still justify itself)*
- ☐ Other ________________

Landscape goals (check all that apply)

- ☐ Beauty
- ☐ Food production
- ☐ Erosion control
- ☐ Slope stabilization
- ☐ Fire break
- ☐ Privacy screen
- ☐ Windbreak
- ☐ Outdoor living
- ☐ Microclimate modification (e.g., windbreak, increased cooling via evapotranspiration, shade)

Water System

Prospective and existing water sources:

- ☐ Well _____ gpm

depth of water table in wet ____, dry season _____

- ☐ Spring _____ gpm (minimum)
- ☐ Meter _____ (size)
- ☐ Rainwater harvesting
- ☐ Runoff harvesting
- ☐ Other ________________

How is your water supply constrained by power supply, economic, ecological, or availability considerations?

Quantity of water ______
Security of water _______
How much does water cost? _______

Volume of onsite water storage _______

What are the water security issues? (E.g., no power = no water = dead fruit trees in a month)

To what degree do you want to or have to conserve?

Existing Wastewater Treatment Facilities

- ☐ Septic: Is it failing?
- ☐ Sewer: Where does it go?
- ☐ Greywater system: Is it functioning satisfactorily (yes/no/sort of)? If not, how?
- ☐ Composting toilet
- ☐ Constructed wetland
- ☐ Other: ______________________________

Special wastewater disposal constraints?

* See definition p. 5.

Population of Water Users

Average population ____
Minimum population _______
Peak population ____
Duration and nature of peak ________
Max continuous days unoccupied during dry season ____
Pending changes in users/use?

Landscape and Irrigation

Native vegetation type(s):

Land use(s), existing and planned:

Irrigated area: Current ______
Potential ______
Existing freshwater consumption?
What is the existing/planned irrigation system?

Is the landscape fenced or free of browsing animals?
Important trees to irrigate?

Slope

Is the area to be irrigated below the greywater source?
Slope % _____
*(Note a Branched Drain system on a 2% slope takes **four times** the labor to build than one on a 4% slope. If the slope is under 2%, it will be very challenging.)*
Are there erosion and/or slope stability (landslide) issues?

Slope aspect (orientation)_____

Soil and Groundwater

Soil type(s): ____________________________
Soil fertility: ____________________________
Digging ease: ____________________________
Permeability (has there been a perk test?)
minutes/in _____ location___________
minutes/in _____ location___________
minutes/in _____ location___________
(Note location(s) and values of perk test onsite map)
Minimum seasonal groundwater depth, seasonal variation:

Distance to nearest year-round surface water____

Distance to nearest seasonal surface water____

Where does runoff go?

Greywater Sources

Fill out table, mark onsite map with quantities of water

Source	Possible to irrigate downhill (y/n)?	Plumbing accessible (y/n)?	Quantity and variability of water, surges, conservation measures, comments
❑ Washing machine			
Bathroom 1			
❑ Bathroom sink			
❑ Bathtub			
❑ Shower			
❑ Toilet water			
Bathroom 2			
❑ Bathroom sink			
❑ Bathtub			
❑ Shower			
❑ Toilet water			
❑ Outdoor shower			
❑ Utility sink			
❑ Wood burning tub			
❑ *Lavadero* (washboard)			
❑ Kitchen sink			
❑ Dishwasher			
❑ R/O water purifier			
❑ Dishwasher			
❑ Water softener flush			
❑ __________			
❑ __________			
❑ __________			

Climate

Annual rainfall _____
Maximum evapotranspiration (inches or cm/week) _____
Minimum evapotranspiration (inches or cm/week) _____
Growing season (frost to frost) _______________
Minimum temperatures _____
Typical max duration w/o significant rain _____
Duration of snow cover ______
Solar exposure (directions) ______
Hours lost from sunrise _____, sunset ____due to surrounding geography and trees
Greenhouse possible? _____ (especially good for cold, wet, low perk locations)

Forces of Nature

Predictable disasters, which may affect the design:

- ☐ Flooding
- ☐ Torrential rain
- ☐ Landslide
- ☐ Fire
- ☐ Very high wind
- ☐ Extreme drought

Users and Maintenance

To what degree are the users interested in understanding/maintaining the system?

Is the system public?

Will there be a person responsible for maintenance?

What are the maintenance goals or constraints?

Regulatory and Social Climate

Will the project be permitted?

Might it be subject to later inspection as part of another project?

What is the applicable greywater code? Other legal considerations?

Neighborhood appropriateness?

Economics

Budget constraints?

Do you own the land where the project is to be built?

How long are you planning to stay there?

Is resale value a concern?

Are there time and money constraints for maintenance, repair, and system replacement?

Is it imperative that the system meet a particular economic payback timetable, or is doing the ecological thing the overriding concern?

Materials and Skilled Labor Availability

Where are plumbing parts and plants coming from?

Are biocompatible cleaners available?

Who is going to do the installation?

Site Map and System Elevations

A ⅛" = 1' scale, 1' contour map[m] **of the site and a ¼" = 1' plan of the structures involved would be ideal,** but any sort of sketch is a help. The map or other description ideally would show topography, property lines, septic tanks, leachlines, wells, surface waters, buildings, major vegetation, and irrigated areas, existing and planned. Aerial photos can help for some sites.

If you're sharing this information with people involved in the project offsite, take snapshots showing general feeling of the site and any special features, indicating the location and the direction of each shot with a letter and arrow on the site map.

Make copies of the map and sketch the possible ways to connect the greywater sources with irrigation/treatment areas.

The elevation relationship between features such as buildings, foundations, walkways, greywater sources, septic or sewer inlet, and irrigated areas is critical.

For all Branched Drain system installations, I strongly suggest making an elevation view drawing as well.

[m]*Metric: 1:50 site map, with 25 cm contours, 1:25 house plan.*

Appendix B: Measuring Elevation and Slope

Naked Eye

The unaided eye is good for determining the direction of a slope. It is, however, amazingly inaccurate for judging the height difference between two points, especially if there is anything but an uninterrupted, even slope between them.

Checking levels with a water level. *Person at right adjusts tube until water level is at her reference mark. Person at left measures from water level in tube to ground.*

Water Level

To measure elevations across short distances up to 30' *(10 m)*, use a simple water level (upper photo). This can be a length of ⅜" *(10 mm)* transparent plastic tubing, or a garden hose with clear plastic extensions. Both are available at hardware stores. Make sure your tubing does not leak. To ensure it is 100% bubble-free, fill it by siphoning water through the tubing until you see no more bubbles zipping through. When moving it, put thumbs over the ends. Raise or lower the line until the water surface at your end is at the elevation of the first point. (If this point is too close to the ground, use a reference point a measured distance above it.) Then have a partner at the other end measure from the water height (which is the same as at your end) to the second point. The difference between the two measurements is the elevation difference between the two points. (There is another version of the water level called the reservoir level, which can be used by one person but is more awkward for greywater work; see figure at right.)

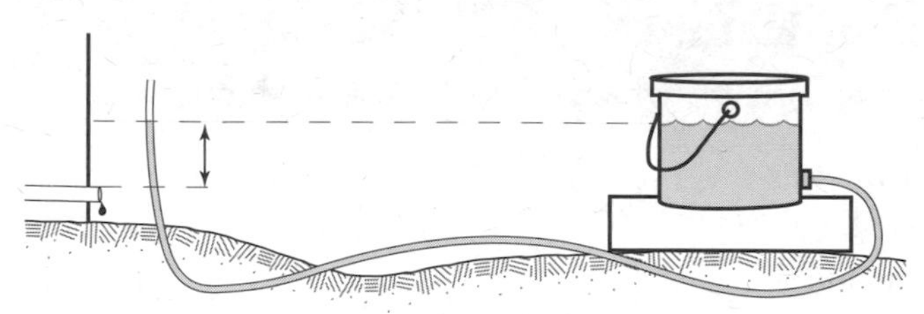

Reservoir water level. *Can be used by one person—water level in bucket barely changes as tube moves up and down. You have to start over if you spill any water, though.*

Checking levels with a water transit. *Person at right adjusts tubing height until her eye and the near and far water levels are exactly in line (don't drink coffee before doing this!) while person at left moves his finger up and down tape measure until his finger is exactly in line as well. Reading on tape measure minus distance from her eyes to ground is height difference between locations.*

Water Transit

For long distances (as far as you can see), you can use a homemade water transit—a 3' *(1 m)* length of transparent tubing filled with water. Hold it in a U-shape with one end near your eye and the other at arm's length, and you can accurately transfer the elevation of your eye to a measuring tape above a distant point or points (lower photo).

There are lots of scenarios for how to measure using this technique. I suggest you make a drawing showing the heights to keep the accounting straight. Here are a few examples:

Sight along the near and far water surfaces in the level to the first point (or to a reference point a measured distance above or below it). Take your time to get the two water levels, your eye, and the point exactly in line. Have a partner mark the reference point. Now turn, keeping your eye at the same height, and look out toward the prospective outlet or obstacle or whatever is at the second point. Have your partner hold a tape measure with the end on the point, and the tape extending up from it vertically. Have them move a finger up and down until it is in line with your eye and the two water levels (lower photo). The difference between your measured eye heights above the two points is their elevation difference.

You can also stand on one of the points and look to the other. Subtract the distance between your eye and the ground from the height of your partner's finger above the second point to get the height difference between the two points.

These measurements can be chained together to go great distances and around visual obstacles with amazing accuracy, if you are careful. Using this technique, I surveyed across a canyon and through trees for the freshwater supply line for the house in Real World Example #3 to within 4" in 1,000' (99.97% accuracy). Maybe we were lucky, but I'm in no rush to buy a pricey "real" transit.

Tubing ends joined with a smaller piece of tubing

+2%

Water level

-2%

Water transit variation that can be used to measure slopes as well as horizontal plane, by sighting to marks above or below water level.

Abney Level

The Abney level is a scope with a split view that shows your distant point on one side and a bubble level on the other. It can measure steep slopes as well as a flat plane. It is especially handy for laying out Branched Drain distribution plumbing, with the slope just set at 2% (though you have to be careful not to space out and sight upstream with the slope set to downstream).

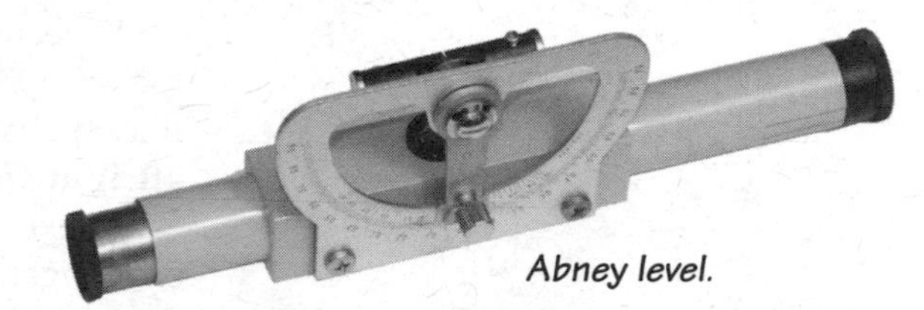

Abney level.

Transit

A modern transit does just about everything: measure a flat or sloping plane, and even distance (by bouncing a laser off a distant object). For this, expect to pay upward of $900.

Transit.

Appendix C: Cold Climate Adaptations

If you live someplace cold, you must evaluate all greywater system designs in light of possible freezing. Knowledgeable local contractors can ensure that proper precautions are taken. For example, all plumbing should drain between uses, be below frost line in the soil, be insulated, or be in a heated space.

The heat in fresh greywater, plus the microbial activity it fuels as it biodegrades, make freezing rare in distribution systems such as box troughs and leaching chambers, even in very cold climates. Heaping leaves or other insulating material over the treatment area helps a great deal. To switch distribution from a shallow, freezing-prone zone to a deeper, frost-free one, use a manual valve, or a passive, automatic bypass (Figure C.1). (In remote applications where human contact is not a concern, frozen greywater has been successfully accumulated on the surface of the soil and treated when it thaws.[3])

FIGURE C.1: AUTOMATIC SWITCHES FROM SHALLOW TO FROST-FREE ZONES / LEVELS

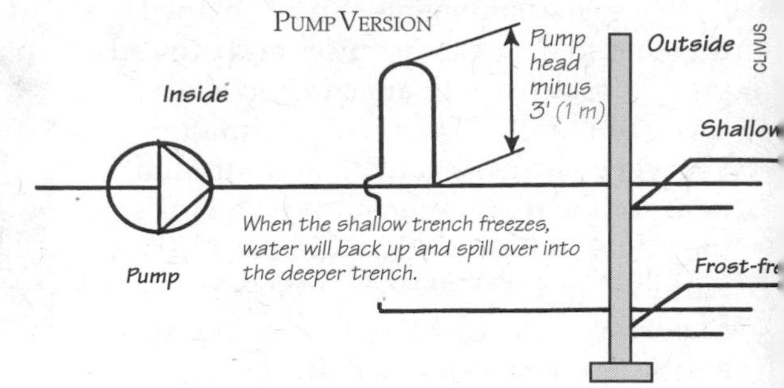

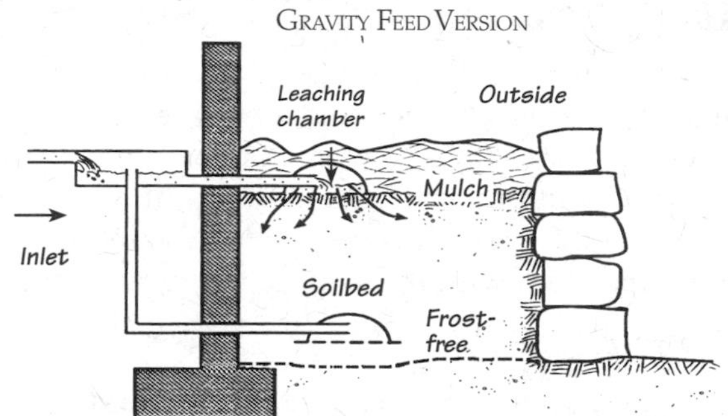

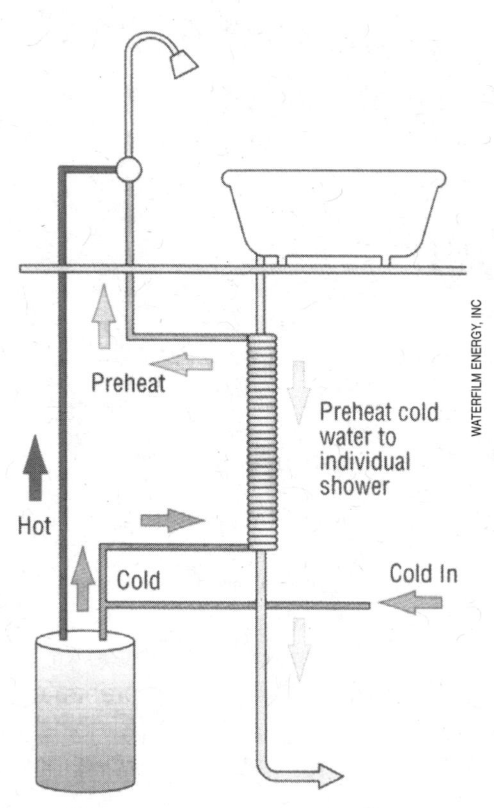

Falling film heat exchanger extracts up to 60% of the heat from greywater and uses it to preheat incoming shower water.

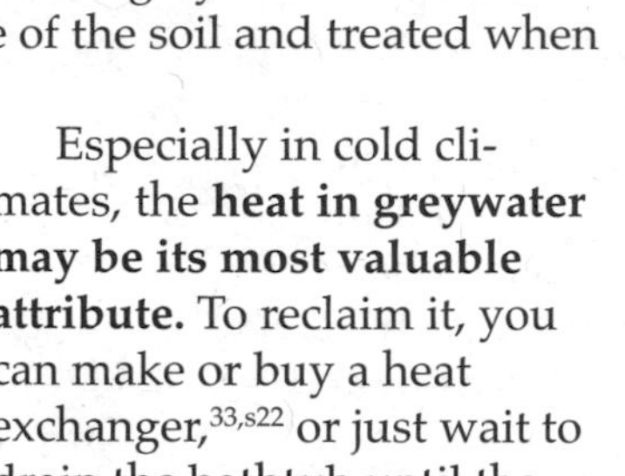

Especially in cold climates, the **heat in greywater may be its most valuable attribute.** To reclaim it, you can make or buy a heat exchanger,[33,s22] or just wait to drain the bathtub until the heat has escaped into the house.

Here is a list of design suggestions for greywater systems in cold conditions:

- **Continuous downhill slope in all collection and distribution plumbing**—No standing water = nothing to freeze. Greywater is warm enough to melt a film of ice in distribution plumbing each time it drains, without buildup. (Branched Drains are a distribution system with continuous downhill slope.)
- **Free flow surface outlets**—At least 6" *(15 cm)* of fall from the greywater outlet to the surface the water lands on will prevent the outlet from plugging with ice. The ground should slope away steeply from the outlet at first, then may gradually flatten—transitioning from, say, 4% slope to 1–2% within a couple paces of the outlet.
- **Insulate the system**—Continuously downhill-sloping pipes may not require any insulation. However, burying them in the soil and covering them with mulch certainly provides some.
  Mulch basins generally stay unfrozen well above the frost line, because the heat from applied greywater and internal composting is held in by insulating mulch and snow. Composting heat can be cranked up by adding plenty of carbonaceous yard waste (leaves, for example) in the fall, then adding nitrogen via greywater (by peeing in the shower, for example) throughout the winter.
- **Apply greywater in a warm microclimate**—The south and west sides of a house, within a windbreak to the north, are significantly less frosty.
- **Apply greywater in a Solar Greenhouse**—A Solar Greywater Greenhouse is the ultimate in favorable microclimates, and the best option for year-round treatment. The treatment capacity per square foot in a greenhouse can easily be many multiples of the capacity outdoors. (Each 18°F *(10°C)* increase in temperature doubles the rate of most biological reactions.)

A snowbound greywater greenhouse *in Massachusetts, filled with fresh veggies.*

Appendix D: Greywater in the Non-Industrialized World

There are special considerations to take into account when working with greywater in the non-industrialized settings in which most of the world's population lives:

- **Water is often scarcer**—Especially where carried in by hand. For this reason, greywater is much more valuable.
- **Greywater irrigated produce can contribute significant nutrition**—Especially when nearby water sources dry up and direct irrigation with freshwater doesn't make sense.
- **Money is often scarce to nonexistent**—So *any* purchase can be a barrier to implementing a greywater system.
- **Infection rates can be high and sanitation norms low.**
- **It is not unusual to have greywater flowing on public streets and in surface waters**—Where it is a health threat.
- **Many villagers prefer to have washing and bathing occur outdoors or in structures separate from living quarters.**
- **Shade in the tropics is a life necessity**—This can be provided by greywater-fed trees, making outside tasks such as hand laundering much more pleasant.
- **People's immune systems are generally stronger**—So a moderate level of pathogen exposure is sustainable.
- **Cleaning products**—Can be better (home-made soap) or worse (commercial powdered laundry detergent with high levels of fillers and non-biodegradable detergents).
- **The concentration of solids in greywater is often higher**—Due to lower per capita water use.
- **Laws are rarely a consideration.**

Landscape Direct, Bucketing, and Drain Out Back systems are all adaptable to these conditions. However, in many circumstances, the system that follows is ideal...

Greywater Furrow Irrigation

Furrow Irrigation is an ideal system for the non-industrialized world. A greywater system reduced to its most elemental form, it consists entirely of soil and plants. It can be owner-built and maintained with nothing more than a shovel and a hoe or rake.

The genesis of this system (for me—I've no doubt it has been independently invented countless times) was the resistance of villagers I work with in Mexico to using mulch. *(Note: This description springs entirely from my experience installing dozens of these systems with village women in Maruata, Michoacán.)*

Traditionally, yards in this coastal village are bare swept earth. With the movement of greywater unimpeded by mulch, furrows in the earth eventually suggested themselves as an option for delivering greywater to plants. Utilizing greywater to irrigate home orchards provides multiple benefits: It eliminates fetid black puddles that otherwise provide habitat for mosquitoes that carry dengue fever, and it can irrigate trees that give fruit and shade, improving nutrition, health, and the economy.

Greywater "main" running down a Tijuana street—*a bona fide health threat. In theory, this water could produce fruit and green relief in a sanitary way. Unfortunately, low rainfall, clay soil, and high salt concentration from hand washing with small amounts of hand-carried water and generous amounts of "Fab One-Shot" renders this greywater unusable. Though it is year-round water in a desert, not even weeds grow from it. An enlightened soap factory with a line of biocompatible cleaners and a suitable marketing plan could dramatically change neighborhoods like this all over Mexico, transforming fetid, mucky streets with thriving shade- and fruit-providing trees.*

Shallow hand-dug well next to a flooding, feces-filled river.

Washwater from a home that feeds a black puddle *like this can be rerouted into irrigation furrows (next page), which water and feed about 20 fruit trees. Because the water is spread over a wide area, it never stays on the surface long, so it doesn't stink or breed vermin.*

Characteristics and Advantages

- **The materials are simple and cheap, and little equipment is necessary**—Soil, trees, a shovel, and a rake or hoe to make the furrows.
- **The system requires only 10 minutes of maintenance per week.**
- **It is necessary to take a multitude of considerations into account to build them optimally**—The design depends on the owners, soil, slope of the land, water sources, sun, existing trees, and more.

Irrigation furrows for bananas. *The more sloped the land, the more sinuous the channel. Note the coconut husk that serves as a valve where the trenches divide.*

Principles of the Design

- **Water supply above**—Water storage and water use areas should be located above the areas to be irrigated.
- **The slope and width of the furrows is varied to release more or less water as needed**—For example, more sloped, narrower trenches near the water source and between plants infiltrate less water, whereas flatter, wider trenches over plant roots infiltrate more water.
- **So the water does not stagnate, it is distributed widely, to sufficient plants to absorb it all**—A property with gravity flow spring water is going to need much more area than a house where water is bucketed in.
- **Systems designed so water does not puddle or soil remain saturated more than 24 hours**—That way, the water maintains a clear grey color without an unpleasant odor. If it remains stagnant longer, it develops a black color and a nasty smell.
- **Each property should infiltrate its own greywater**—And not allow it to drain into neighboring properties or into the public road.
- **The furrows are to be adjusted so the water is distributed uniformly.**
- **You can distribute water over more area more easily if the greywater sources are separated**—For example, one spot for dishwashing, another for clothes washing, another for bathing, rather than all these activities taking place in one very wet area and the rest of the land being dry. Other factors to consider when locating water fixtures and work areas are privacy, sun, soil quality, and protecting natural waters from contamination.

Tools and Materials

- **Tools**—Shovel or stick, rake or hoe
- **Optional tools**—Wheelbarrow, pick, bucket
- **Materials**—Soil, fruit trees
- **Optional materials**—Compost (so the plants grow well), stones, bricks, cement (to pave bathing/washing areas where the water falls), clay (to line furrows to make water infiltrate more slowly)

Implementation

- **Pave the areas where water falls**—Sometimes falling water excavates a hole in the ground, which then forms a stagnant puddle right next to where water is used. By paving it with stone, bricks, cement, or rot-resistant wood, you can avoid this.
- **Raise the areas where people walk or stand to wash**—In those areas, the soil tends to compact and erode, leaving the ground lower and wetter. It is healthier and more comfortable if these places are higher and drier. This problem can be fixed by laying down extra dirt once in a while, or paving these areas.

Garden committee members installing a masonry floor *(normally men's work) over the puddle that a neighbor used to stand in while doing laundry.*

- **Review the levels**—Where can the greywater flow by gravity? If necessary, this can be verified by using a water level. Normally, ⅛"/ft *(1%)* slope is necessary for the water to run down the furrows.
- **Review where runoff flows**—The ideal situation is that the furrows for rainwater drainage are not the same as those for greywater drainage. This is because heavy rain can whisk greywater off the property without treatment.
- **Connect the sources of greywater with irrigation need**—Once you've settled where the greywater comes from and how much there is, you can design a connecting system of furrows. The furrow routes can be scratched in the dirt as you work them out.
- **Decide how many furrows**—If there is a large quantity of water, make at least two systems of furrows. That way you can change sides when necessary, allowing the other side to rest. The "valve" can be a simple wye in the furrow, with a coconut shell or stone to close off the side that is not in use. Another technique is a movable channel or hose at the water source, shifted from one furrow to another. It is preferable to have this first "valve" very close to the water source, ideally in cement or pavement. This is because the beginning of the furrow always has more water, and thus needs a rest more. Moreover, the watercourse before the valve never has a rest, so it should be either very short or of an impermeable material.

Form the Water Channels

If you want more water to infiltrate, for example, near the plants, or in not very permeable soil, make the channel:

- Wider
- Flatter
- Longer and more curvaceous
- More permeable
- With basins to one side or in the middle where water will pool; if they are in the middle, they have to fill up before the flow continues—a technique that is only useful when the land is at a steeper incline and there is a lot of water

If you want less water to infiltrate, for example, where there are no roots, in very permeable soil, or in very flat soil, make the channel:

- Narrower, especially in the bottom of the channel where the water actually runs
- Steeper
- Narrower and straighter
- Less permeable; this can be done by smearing the channel with clay, to form a half-tube of clay—remember that greywater itself will lower the permeability of new furrows over time

Typically, the channels will be straighter, narrower, and steeper near the water source, and flatter, wider, and more winding farther away.

For most trees, the roots grow farther from the trunk than the branches do. It is only necessary to make channels pass near the trunk for newly planted trees.

The best situation is that the channels and basins have enough volume to absorb all of the water.

If the volume of water changes dramatically, as it does in tourist lodging sites, the trench cross-section can be narrow at the very bottom, and wide above.

Planting coconuts for shower runoff.
Hiring women gets around the problem that men often spend most of the money they earn on beer. Paying women the same as men (instead of the going rate of about half) helps advance social equality.

Plant the Plants

Plant the trees where they can take best advantage of the greywater and provide greatest benefits of shade, privacy, windbreak, etc. Typically, plants that require a lot of water, such as bananas, will be near the water source, and plants that need little water will be nearer the end of the furrows.

Plant intensively right around areas used for washing/bathing. These plants will form a dense network of roots below these constantly wet areas, helping to dry them, as well as providing shade and privacy above. In the tropics most work areas require shade. This system creates low cost, fresh green shade that gives fruit and deepens over time, instead of costly, high maintenance shade roofs made from cut trees that deteriorate and provide less shade with time. Also, as the water evapotranspires, it cools the air, making the microclimate more comfortable.

A 10 lb (4 kg) papaya from greywater irrigated tree at a house with only hand-carried water for domestic use. The papayas which weren't near greywater trenches didn't bear fruit at all.

Test the System

Pour water into the channels to see where it goes. Adjust anything that needs adjusting.

Explain the System and its Maintenance

Explain to the owners how to maintain the system. Explain to the children that it is fine to play with the flow of water as it is being sent to the plants, but that they should use a stick, hoe, rake, or their feet, not their hands, because the water is dirty. Also, it is necessary to wash their hands immediately afterward if they do touch the water.

Review How the System Works over Time and the Change of Seasons

Return to the system in a week, a month, and with the change of seasons (dry to wet or wet to dry) to see how the plants are growing and how the maintenance is going.

Example #6: No-Money Ecotourism—Michoacán (Greywater Furrow Irrigation)

This particular Furrow Irrigation system can handle high season shower water shock loads 20 times the normal loading rate, in a tiny area.

System: Greywater Furrow Irrigation, chosen due to low cost, low ecological impact, and community's aversion to mulch.

Context: Maruata, an Indian village on the Pacific coast of Mexico, at 18° latitude. The owners of this site—the first where I tried this system—are well off for the region but had no tools to speak of and not enough money for even one length of pipe.

Goals: Get greywater out of the river and public streets in a way and for a cost such that the locals would maintain it, and replicate it on their own for other sites.

Design and installation issues: The main issues at this site are its tiny area (250 ft^2, *25 m^2*) and huge shock loads. The system is designed to handle ten\ days of 200 showers a day during *semana santa* (holy week). One time another huge group unexpectedly showed up right after holy week. With the ground already saturated, they hit it with another 800 showers in four days—I hear it smelled pretty bad then, but otherwise it handled it.

In contrast, the normal use level is about ten showers a day. The trenches have a special shape to accommodate this.

Halfway through construction the owners mentioned that during cyclones (every 1–4 years) the river rises and the entire valley, including the bathhouse and greywater infiltration/evapotranspiration bed, is inundated by thigh-deep, debris-laden water. Also, the cyclone winds strip everything aboveground bare of leaves. The design is adapted to this. After a cyclone, the swales can be reformed, and can be re-planted if plants don't re-sprout. The coconuts make new leaves and the bananas and papayas sprout back up again from their roots. (Even the regular annual monsoon is extreme. My daughter and I collected dry-season water samples one day in June and the wet season started that night—with 8" *(240 mm)* of rain in eight hours.)

Costs and benefits: Initially grading and forming the basins took a number of hours. The biggest cost was my design and construction help, which was paid for by an American philanthropist. (Once there was an example to copy, this was not a factor for subsequent systems.) The second biggest cost was fencing to keep roaming pigs out. This is critical, as they root up the soil and eat every living plant. Fencing is expensive, and a wastewater treatment area has low economic productivity compared to fenced camping space for travelers. Pig-proof plants/living fence would be a breakthrough for greywater treatment for this area. *(Note: The villagers banned free-roaming pigs a few years later, and gardens are now spreading outside fences all over the village.)*

Figure D.1: Public Shower Furrow / Infiltration Basin in Michoacán

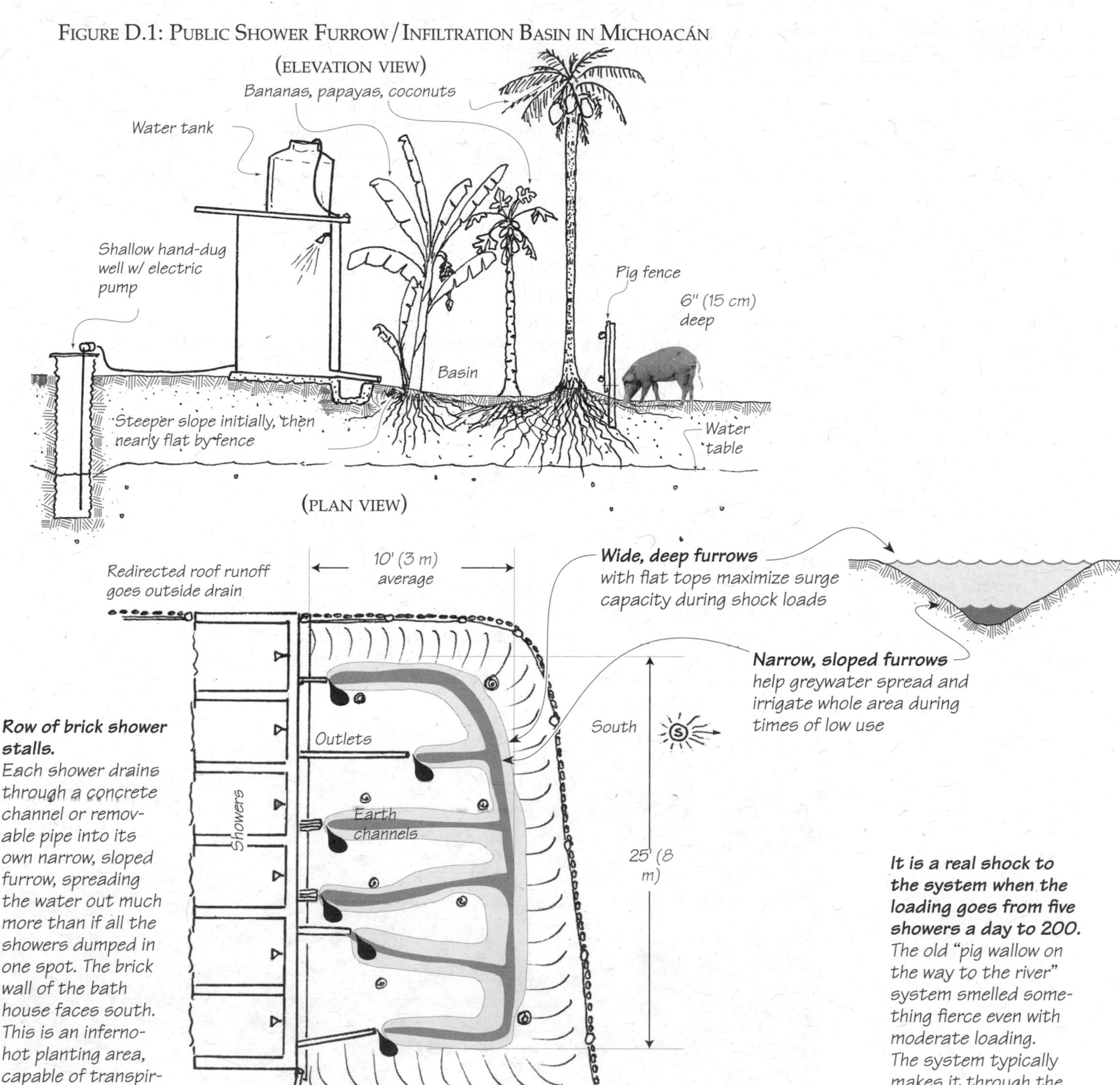

Row of brick shower stalls. Each shower drains through a concrete channel or removable pipe into its own narrow, sloped furrow, spreading the water out much more than if all the showers dumped in one spot. The brick wall of the bath house faces south. This is an inferno-hot planting area, capable of transpiring a LOT of water.

Narrow trench bottoms wet during moderate loading.

Basins full to the brim at the peak season.

It is a real shock to the system when the loading goes from five showers a day to 200. The old "pig wallow on the way to the river" system smelled something fierce even with moderate loading. The system typically makes it through the peak season without any anaerobic odor at all (though just barely, I suspect). Basins progressively fill during the day, then soak in overnight. The standing water would subside all the way by about 5 am, then get refilled about 6 am when people started taking showers. This system pushes loading rate about as far as you can push it.

In the first iteration of the system, the basins were filled with mulch. Maintaining the mulch level in the face of tropical decomposition rates would have been easier if the owners separated mulch from their existing waste stream and added it to the mulch basin instead of burning it with everything else. Also, the locals prefer bare swept ground, and the owners perceived the mulch in the basins as unsightly. What's more, they were concerned that it would breed scorpions and snakes. This fear was probably exaggerated—considering the hundreds of scorpions already overhead in the *palapa* roof, who cares about a few behind the bathhouse?

Overall I'd call this system a clear technical success, and a marginal social one. The owners were maintaining the plants beautifully when last I saw it, and had actually extended the fenced area. However, they were still not getting the hang of maintaining the form of the basins to accommodate both low and peak flows, which is very tricky. On the other hand, I was pleasantly surprised to see other villagers dedicating resources to making more such systems. Production of coconuts, bananas, and papayas in the planter bed helps tip the scales in favor of maintaining the system.

Funder's view: *"Kudos for design, execution, and longevity of system. That it was not torn up, dismantled, and junked right off speaks volumes. I have noticed other works by 'furriners' are generally given short shrift by Maruatans."*

Builder's view: *"I learned how to take the greywater that is always generated from houses and use it for irrigating fruit trees. The water is routed to the whole orchard by means of small channels in the earth. By the* pilas *(water tanks) where we wash, it is now cool and shady with trees that are covered with fruit. Also, it is more healthy, as before the water was all over the place, and mostly ended up in nasty puddles that bred mosquitoes."*

Owner's view: *"A group of us that own tourist facilities on the beach identified greywater in the river as one of the problems we'd like to fix. My* palapa *is the farthest upriver, so getting the greywater out added immediately to the length it runs clean, which feels good. Also, it smells better, and we like the fruit."*

Jardineras replacing dead shade (plastic tarp) over a dishwash station with live shade *(coconut trees from their cooperative nursery, in back of bicycle). In five years the trees were twice as tall as the rock. They give deep, cool shade—not to mention building material, cocos, and protection against erosion.*

Team of* Albanilas *(female masons) making a cement channel *to route greywater from a washing area to the first split between Greywater Furrows.*

Appendix E: Pumps, Filters, and Disinfection

Pumping, filtration, or disinfection are complications you're better off without unless there is no way around them. Gravity driven systems are much simpler, cheaper, and less failure prone. In certain contexts, however, these things make sense, so this section explains how to do them with minimal trouble.

Pumps

The ecological cost of pumps—their manufacture and power consumption—is considerable. There are only two justifications for pumping:

1. When the only place the greywater can go is uphill
2. When the improved irrigation efficiency from distributing greywater more widely and evenly is worth more than the materials and electricity involved in pumping

A pump will likely be the most expensive part of the system. But don't be tempted to buy a cheap pump. You will end up buying the good one anyway, after the cheap one fails. Don't buy a pump unless it is rated to handle at least the lift and volume your system requires.

An effluent pump that is rated to pump a ¾" solid won't choke on unfiltered greywater and is thus the most reliable. These are not cheap: $150 for a low pressure one, $500 for a high pressure one (2005). They can be used to pump unfiltered greywater to the landscape through plumbing that has orifices ½" *(12 mm)* or larger, or through a rapid sand filter. They live submerged in a surge tank and require a float switch to turn them on and off at the appropriate water levels.

Throwaway, cheap, submersible sump pumps are, unfortunately, the most commonly used greywater pumps. They require a pre-pump filter and a backup filter. Experience has shown that filters are not maintained for long, so the pumps are doomed to fail (see Filters, following). They are really only appropriate for temporary, drought emergency systems.

A mercury float switch at the end of an electric cord controls many pumps. An advantage of mercury float switches is that the water level at which they turn on and off can be adjusted by tethering them with a different length of free cord. They can also be used with any pump. The disadvantage of this type of float switch is that it can hang up on the pump or the side of the tank and fail to turn on or, worse, fail to turn off. If not held in place by stiff outlet plumbing or such, the pump tends to migrate around the bottom of the tank until it finds the spot that keeps the float switch from shutting off. Sump pumps require submersion for cooling, so running dry will fry them. They are relatively unaffected by pushing for hours against a clogged distribution line, as long as they are submerged—they just turn into water heaters.

An integral float switch such as the ones on the top and bottom pumps pictured at right has the advantage that it is relatively unlikely to hang up on anything. It turns on in 7–10" of water and shuts off in 2–4". This can lead to a small amount of water being pumped out frequently, which could mean that only the first plant or two gets watered.

Pressure sensors and other schemes for turning pumps on and off often clog with greywater.

Beside being expensive and a maintenance hassle, pumps use electricity. In an average home, *a greywater pump can easily be in the top ten energy consuming appliances, accounting for 3–10% of total household consumption.* To maximize efficiency:

- **Minimize the distance that water drops in collection plumbing and surge tanks before reaching the pump**—A wide, flat surge tank wastes half the fall of a tall, narrow one. Reducing the drop by half may not be that hard and significantly reduces electricity consumption.
- **Don't pump water any higher than necessary**—If you only have enough greywater to water half of your landscape, greywater the lower half. Unlike freshwater systems, it rarely makes sense to pump greywater to an elevated tank and gravity feed from there.

An effluent pump that doesn't need filtration and will last for many years is the way to go (a $500 Zabel p-se-12t pictured).[s5]

A cheaper effluent pump doesn't provide quite the longevity or pressure (a $160 Little Giant Model 8E CIA RFS pictured).[s8]

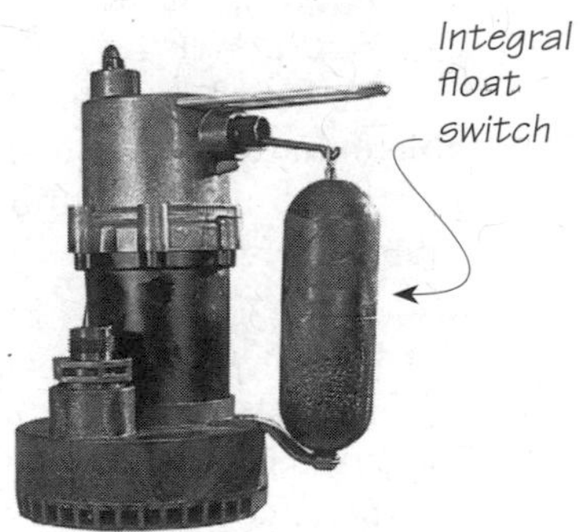

One of the better throwaway pumps that require a filter (a $200 Little Giant Model 5.5-ASP pictured).

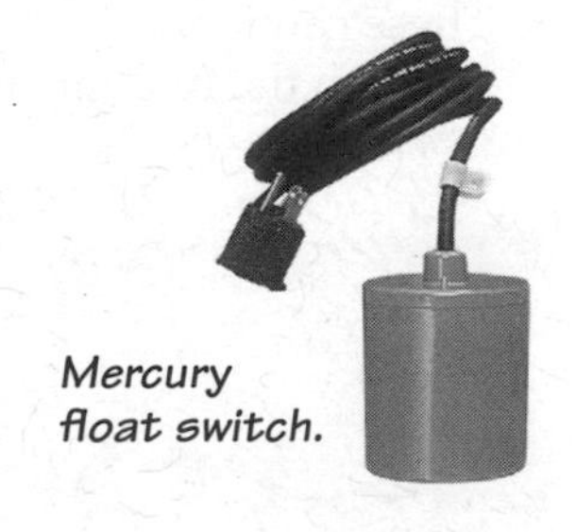

Mercury float switch.

- **Use adequate-size distribution pipe to minimize friction loss**—1" pipe should be adequate for a home system's pressurized greywater main.
- **Size your distribution system for adequate flow, and don't pressurize it any more than necessary**—If all the distribution orifices together can only handle half the flow the pump puts out, the pump has to run twice as long. Likewise with pressure; more than 10–15 psi *(70–100 kPa)* is not necessary for most systems, so a pump that maintains 30 psi *(200 kPa)* uses two to three times the necessary energy.

Filters

Filtration is the biggest problem in the systems that require it. Hair and bacterial slime are particularly difficult to handle. Yard-long hairs can work their way through a fine-mesh filter and then wrap around the pump rotor. Bacterial slime can grow on the "filtered" side of the filter, then slough off in large chunks. Filter cleaning is a disgusting task. This is the #1 cause of abandonment for any system that requires filter cleaning. A cheap pump won't work without a pre-pump filter, followed by a net filter around the pump itself. The occasional extremist will clean pre-pump filters, but *no one* cleans the around-the-pump backup filter... with the result that the cheap pump dies.

Filtration is definitely an area for more research. Here are the current options:

No Filtration

If at all possible, no filtration is definitely the way to go.

Solids are easily handled by collection plumbing and are not a problem for soil—they're just compost. A system without a pump or any passages smaller than 1" *(2.5 cm)* doesn't require filtration. Expensive effluent pumps can handle solids, but even these may occasionally have problems with hair. It is all but impossible to achieve automated, even distribution of greywater without filtration (except to a limited extent with a Branched Drain system or an effluent pump to open outlets), but the absence of filter maintenance weighs heavily in the trade-off.

Pre-Drain Filtration

If your system requires only minimal filtration, **catch the solids before they get into the pipes** and get grosser. A simple mesh screen over the kitchen sink drain, for example, filters out a large proportion of solids.

Net Bag

The net bag is the most common, but not especially recommended, greywater filter (see Figure 8.1, photo at right).

The net can be pantyhose or a fine-mesh filter bag for airless paint sprayers (available in hardware stores). Mesh bags of 75-micron were used as pre-filtration in drip irrigation, but this technology fell into disfavor due to an excessively short filter-cleaning interval. Net bags are disgusting to clean. Pantyhose are readily available and cheap enough to throw away when full, especially if already worn. They also stretch to shed the aforementioned bacterial slime.

Net bag filter in a Drum with Pump.

Net bags are notorious for dumping gallons of backed-up crud irretrievably into the surge tank when removed for cleaning (see story under Error: Greywater to Inexpensive Drip Irrigation, Chapter 11). A second filter around the pump is recommended to catch this crud, and a filter on the outlet side is essential with medium sized emitters. Small-orifice emitters would be a mistake.

Pressure Sand Filtration

Successful greywater-to-drip-irrigation systems employ a swimming-pool-type pressure sand filter that automatically backflushes, providing a high level of "hands-off" filtration. These can work well, but are expensive and require lots of electricity to push the water through the filter at 40 psi or so, or the energy equivalent of raising the water 100 vertical feet *(3 atm, or 30 m)*.

Note: A non-backflushing, unpressurized sand filter clogs too quickly with greywater.

Settling (Septic) Tank

These are excellent filters, and the required maintenance interval is measured in years or decades instead of days or weeks.

In a septic tank, water sits still long enough for scum to float to the top and sludge to settle to the bottom, leaving smelly, infectious, but clear water in between. The settling tank should have capacity for 1–30 days' flow. More is better, within reason.

Most greywater systems avoid the worst odor and bacteria proliferation problems by distributing water as it's generated, eliminating settling as an option. If effluent disposal is completely subsurface, storage is not an issue, and this shouldn't matter. Perhaps settling will become more popular in the future, due to its infrequent maintenance. Sludge can be pumped out of small settling tanks with a wet/dry vac.

Septic Tank and Media Filter

This is another proven technique for getting greywater clean enough to use in subsurface drip irrigation (see end of Chapter 8).

Constructed Wetland as Filter

A Constructed Wetland with successively smaller media in the direction of water flow (rocks to gravel) could be a good filter, if sized adequately (see end of Chapter 8).

Caution: Unproven design; Constructed Wetlands are experimental as filters.

Disinfection

Ozone and UV disinfection have both been used successfully. I'd trust ozone more. But like filtration, disinfection is best avoided. Unless the greywater is sprayed in sprinklers (illegal), used in a pond, or returned to fixtures for reuse, disinfection should not be necessary. Plus, disinfection won't accomplish much unless the water is mechanically filtered first. Indoor reuse in fixtures is the most legitimate application for disinfection (see The Household Water Cascade in Appendix F, and Resources[34,35]).

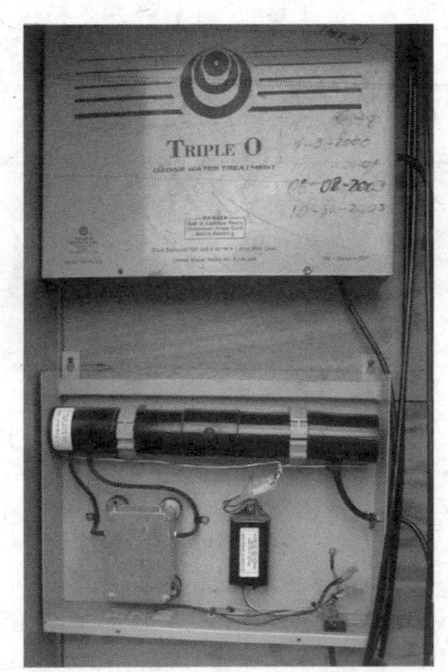

Ozonators *provide sterilization with far fewer environmental and health side effects than chlorine and may be more reliable than UV due to low water clarity in greywater.*

Appendix F: Related Aspects of Sustainable Water Use

Natural Purification

Ideally, all water should cascade through the house and yard by gravity, from the highest vertical level and degree of purity to biological treatment in the soil at the lowest level.[34,35] The result is a house surrounded by an oasis of biological productivity nourished by the flow of water and nutrients from the home.

In the natural water cycle, water is purified in two ways: 1) distillation, where evaporation from the ocean leaves particles and dissolved salts behind; and 2) biological land treatment, where microorganisms in topsoil degrade biological contaminants into nutrients, which are removed by plant roots before groundwater recharge (see p. 18 for effectiveness factors).

Both natural distillation and biological land treatment compare favorably in effectiveness with artificial options such as activated sludge treatment. Conventional water use and treatment, by ignoring the structure and logic of the natural water cycle, encourages excess, depletes and contaminates aquifers, contaminates aquatic ecosystems, and bleeds the land of nutrients.

Natural Purification by Soil Bacteria and Plant Roots

In one cubic foot of topsoil there are:

- *One and a half million square feet of treatment area*
- *Three trillion beneficial bacteria*
- *Enough root hairs to wrap around the perimeter of the US*
- *Countless specialized proteins which pump specific nutrient molecules (e.g., nitrate) through root hair cell walls*

Effectiveness:

- *Slow passage through one foot of soil removes about 90% of pathogens*
- *Code allows untreated drinking water well 100′ from a septic leachfield*

FIGURE F.1: A HOUSEHOLD WATER CASCADE
(THE WIDTH OF THE LINES CORRESPONDS APPROXIMATELY TO THE VOLUME OF WATER FLOWING THROUGH THAT PATH)

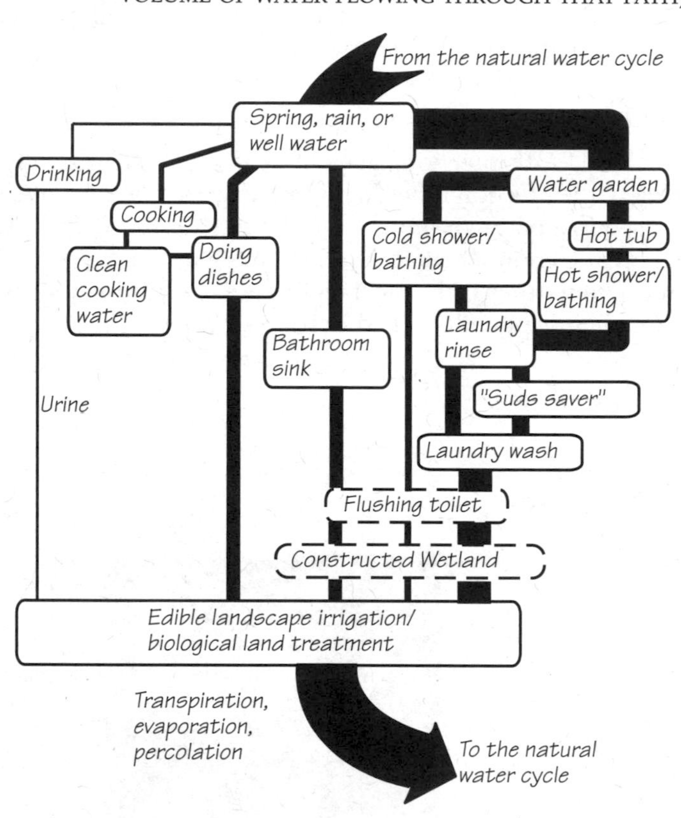

The Household Water Cascade

Water conservation is popular during a water shortage. However, there are more reasons to cascade water and use it ultimately for irrigation: conserving nutrients, conserving the energy used to heat water, and reducing disposal problems. Keep these factors in mind when designing the water flow through a living environment in which water is plentiful.

Elevation relationships are a major consideration in the siting of different elements of the water cascade, such as water tanks, gardens, solar heaters, buildings, cooking areas, kitchen gardens, washing areas, bathing gardens, and edible landscaping. Within the home, drinking is the highest use, followed by cooking, bathing, hand washing, dishwashing, and clothes washing. Ecologically, it only makes sense to dump feces into water when dry feces disposal is infeasible. If so, this use comes last in the cascade.

Even if you're not building your own radical eco-home, cascading can still be employed in simple ways. For instance, water used to steam vegetables or boil pasta can be substituted for some of the hot water for doing dishes. The water might be a little cloudy, but no more than it would be after washing a few dishes in clean water. "Suds-saver" washing machines are a classic example of cascading; laundry rinse water is pumped to a sink or tank, then drawn back into the machine for use as wash water in the next load.

When showering in bathrooms with a shower/tub combination, save the shower water in the tub for hand washing clothes. The water will already be hot, so just add a little detergent. When finished, siphon the water outside.

The "lid": a super easy, inexpensive cascade retrofit *that enables hands to be washed in the flow of freshwater on its way to the toilet bowl.*

Rainwater Harvesting

Even in dry climates, quite a bit of water falls from the sky. In industrialized nations the propensity is to seal the surface of the soil with pavement, roofing, plastic sheeting, and compaction—transforming groundwater recharge zones into flash flood generation zones. (Los Angeles supposedly absorbs only 10% of the water it used to.) In arid lands, water is best stored in covered tanks or in groundwater. Groundwater can be supplemented in wet decades and drawn upon in dry ones, a practice called "conjunctive use." To capture runoff and get it to seep quickly into the ground, spread mulch, add plants, and shape catchment basins and swales in the surface of the soil (see Topography and Mulch, end of Chapter 5).

More significant than rainwater quantity is its quality. When irrigation greywater evaporates, salt is left behind in ever-increasing concentrations. Rain has virtually no dissolved salts and is highly effective for flushing excess salts from the root zone (see Maintaining Soil Quality, Chapter 5). Our book *Water Storage*[10] describes how to store rainwater in soil and groundwater and how to collect it in runoff ponds or tanks.

Rainwater harvesting *provides all domestic water by gravity for a house near Hilo, HI.*

Composting Toilets

Used in conjunction with greywater systems, composting toilets promise a household that uses little water, recycles all of its nutrients and water, and employs no conventional treatment systems at all. Proceed with caution in pursuit of this lofty ideal. Though some composting toilets have worked well for decades, we have collected mixed reviews of their function. Carefully research your toilet decision[36,37,38,s15] (and please share your experiences, as we are researching books on flush toilet alternatives and rainwater harvesting[39,40]).

There are dozens of waterless toilet technologies to choose from for different population densities and contexts.
The Chesapeake Bay Foundation's 32,000 ft² headquarters (above) features capacity for 160,000 toilet uses a year, right on the bay. This is the first LEED Platinum-Certified building. It uses photovoltaics, solar hot water, rainwater collection, natural lighting/ventilation, and ground-source heating/cooling. Its water conservation technologies include waterless Clivus composting toilets (seat and basement treatment units at right). The use of composting toilets demonstrates a key technology for eliminating nutrient pollution of water resources.

Appendix G: Greywater Regulation Revolution

A move is afoot toward sensible greywater laws in the Southwest US. The greywater law in Arizona, the pioneer, is reproduced here. New Mexico has a similar law, Texas has one that needs some work, and Nevada is about to introduce one like Arizona's. In Arizona your greywater system is automatically permitted if it meets these handful of (mostly) reasonable requirements. If only other jurisdictions were so levelheaded! If you have to deal with greywater laws anywhere else, see our *Builder's Greywater Guide*.[6] It has much more information about greywater laws (including how to improve them) as well as other aspects of incorporating greywater systems in new construction and remodeling. (There is also information on greywater laws in our Greywater Policy Center, oasisdesign.net/greywater/law.)

Arizona Greywater Law

R18-9-711. Type 1 Reclaimed Water General Permit for Gray Water

A. A Type 1 Reclaimed Water General Permit allows private residential direct reuse of gray water for a flow of less than 400 gallons per day if all the following conditions are met:

1. Human contact with gray water and soil irrigated by gray water is avoided;
2. Gray water originating from the residence is used and contained within the property boundary for household gardening, composting, lawn watering, or landscape irrigation;
3. Surface application of gray water is not used for irrigation of food plants, except for citrus and nut trees; *[This is a funny provision; it should prohibit direct contact between greywater and consumable portions of food plants.]*
4. The gray water does not contain hazardous chemicals derived from activities such as cleaning car parts, washing greasy or oily rags, or disposing of waste solutions from home photo labs or similar hobbyist or home occupational activities;
5. The application of gray water is managed to minimize standing water on the surface;
6. The gray water system is constructed so that if blockage, plugging, or backup of the system occurs, gray water can be directed into the sewage collection system or onsite wastewater treatment and disposal system, as applicable. The gray water system may include a means of filtration to reduce plugging and extend system lifetime;
7. Any gray water storage tank is covered to restrict access and to eliminate habitat for mosquitoes or other vectors;
8. The gray water system is sited outside of a floodway;
9. The gray water system is operated to maintain a minimum vertical separation distance of at least five feet from the point of gray water application to the top of the seasonally high groundwater table;
10. For residences using an onsite wastewater treatment facility for black water treatment and disposal, the use of a gray water system does not change the design, capacity, or reserve area requirements for the onsite wastewater treatment facility at the residence, and ensures that the facility can handle the combined black water and gray water flow if the gray water system fails or is not fully used;
11. Any pressure piping used in a gray water system that may be susceptible to cross connection with a potable water system clearly indicates that the piping does not carry potable water;
12. Gray water applied by surface irrigation does not contain water used to wash diapers or similarly soiled or infectious garments unless the gray water is disinfected before irrigation; and
13. Surface irrigation by gray water is only by flood or drip irrigation.

B. Prohibitions. The following are prohibited:

1. Gray water use for purposes other than irrigation, and
2. Spray irrigation.

C. Towns, cities, or counties may further limit the use of gray water described in this Section by rule or ordinance.

Appendix H: Measurements and Conversions

How units are dealt with in this book:

Any measurement clearly expressible without numbers or units is expressed without them (e.g., "an arm's length").

Where the text flow is too chopped up by non-essential numbers, they are relegated to footnotes. Metric conversions of pipe sizes are here, only.

Units that are approximations or don't really matter are given to the nearest round number or conventional expression. Thus, a photo of a surge tank might be captioned "55 gal *(200 L)* tank" rather than "55 gal *(206.8 L)* tank."

Using the tables:

Everything in the same row is equal. For example: 1' = 12" = 0.3 m = 30 cm.

Length/Height

Feet ft'	Inches in"	Meters m	Centimeters cm	
1	12	0.30	30	Long as...your foot
0.08	1	0.025	2.5	Long as 1st knuckle of the thumb
3.28	39.37	1	100	A long stride

Pipe Sizes

US	Metric
1/2"	15 mm
3/4"	20 mm
1"	25 mm
1 1/4"	32 mm
1 1/2"	40 mm
2"	50 mm
2 1/2"	65 mm
3"	80 mm
4"	100 mm
6"	150 mm
12"	300 mm

Area

Square meters m^2	Square feet ft^2	Acres ac	Hectares ha
1	10.76		
0.09	1		
4,047	43,560	1	0.405
10,000	107,639	2.47	1

Volume

Cubic meters m^3	Liters L	Gallons gal	Cubic feet ft^3	
1	1000	264.17	35.3	A cube 40" on a side
0.001	1	0.26	0.0353	
0.00387	3.78	1	0.134	1 US gal = 0.833 Imperial gal
0.0283	28.3	7.5	1	
0.2	200	55	7.06	Big plastic or steel drum
113	113,562	30,000	4,010	Standard suburban swimming pool
1,233	1,233,482	325,851	43,560	1 acre foot (af)
0.00235	2.35	0.62	0.0829	1" of water on 1 ft^2
	1			1 mm of water on 1 m^2

Flow

An industrialized world household of four uses roughly 1 m^3 a day of water.
A non-industrialized world household of ten uses roughly 1 m^3 a day of water.
Abbreviations: Gallons per day (**gpd**), Liters per day (**lpd**), Gallons per minute (**gpm**), Liters per minute (**lpm**), Cubic meters per day (**m^3/day**), Acre-feet per year (**afy**)

gpm	ft^3/sec	lpm	m^3/day	afy
1		3.79	5.45	1.61
448	1	1,699	2,446	723
0.26	0.00	1	1.44	0.43
0.183		0.694	1	0.296
0.620	0.00138	2.34	3.38	1

Pressure

Abbreviations: Atmospheres (atm), Pounds per square inch (psi), Kilopascals (kPa)

atm	psi	kPa	Feet	Meters	
1	14.70	101	33.9	10.3	A 1" x 1" column of air from the Earth to space in height weighs 14.7 pounds.
0.0680	1	6.89	2.31	0.703	
0.0099	0.145	1	0.334	0.102	
0.0295	0.433	2.99	1	0.305	
0.0978	1.42	9.80	3.28	1	
6.80	100	689	231	70.3	Max pressure for household plumbing, ideal pressure for fire hoses
1.70	25	172	57.7	17.6	Min pressure for washer, demand heater valves

Resources

Further Reading and Resources

1. **Relative Health Hazard of Different Waters** oasisdesign.net/greywater/law/FecalLevels.xls *Spreadsheet comparing measured and calculated fecal coliform levels. It appears that indicator organisms multiply in greywater plumbing, but there is no indication that pathogens do.*
2. **Sewers, Sludge Treatment, Sludge** www.riles.org/library.htm *Various writings on the evils of sewers and advantages of nutrient reclamation.* oasisdesign.net/wastewater/sewer/beach.htm *A letter about misinterpretation of a study of contamination from oceanfront septics, and the inadvisability of sewering.*
3. **Green Land—Clean Streams: The Beneficial Use of Waste Water Through Land Treatment R Michael Stevens,** Temple University's Center for the Study of Federalism, 1972. 330 pages. *Analysis of large scale land treatment facilities. Currently out of print; there is a copy at the library at Temple U. Their summary table of treatment effectiveness, which is amazingly high, even for overland flow, is reprinted in our* Builder's Greywater Guide.[6]
4. **Greywater Policy Center** oasisdesign.net/greywater/law *Includes California, Arizona, New Mexico, and other greywater laws, and suggested policy improvements.*
5. **Greywater Pilot Project Final Report** Los Angeles Department of Water Reclamation, 1992. Los Angeles Department of Water and Power, PO Box 1111, Rm. 1315, Los Angeles, CA 90051-0100. Phone: 800-342-5397, 213-481-5411, ccenter@ladwp.com www.ladwp.com *First quantitative field testing of greywater health safety.*
6. **Builder's Greywater Guide** Art Ludwig, 2006. Oasis Design, Santa Barbara, CA. Phone: 805-967-9956, Fax: 805-967-3229. oasisdesign.net/greywater/buildersguide. *Supplement to* Create an Oasis *with information of interest if your project is going to involve permitting, inspection, building a system for someone else, writing or applying codes, or greywater research.*
7. **National Small Flows Clearinghouse** West Virginia University, PO Box 6064, Morgantown, WV 26506-6064. Phone: 800-624-8301, 304-293-4191, Fax: 304-293-3161 www.nesc.wvu.edu/nsfc/nsfc_index.htm *Clearinghouse organization with a wealth of information on small to large scale wastewater treatment.*
8. **The Septic System Owner's Manual** Lloyd Kahn, Blair Allen, and Julie Jones, 2000. Shelter Publications Inc., PO Box 279, Bolinas, CA 94924. Phone: 415-868-0280, Fax: 415-868-9053, shelter@shelterpub.com www.shelterpub.com *Everything you need to know about maintaining a septic system. Includes section on alternative systems.*
9. **Rainwater Harvesting for Drylands and Beyond** Brad Lancaster, 2006. Rainsource Press, 813 N. 9th Ave. Tucson, AZ 85705. brad@harvestingrainwater.com www.harvestingrainwater.com *Available from* oasisdesign.net/catalog
10. **Water Storage** Art Ludwig, 2009. Oasis Design, 5 San Marcos Trout Club, Santa Barbara, CA 93105-9726. Phone: 805-967-9956, Fax: 805-967-3229 oasis@oasisdesign.net oasisdesign.net/water/storage *Tanks, cisterns, aquifers, and ponds for domestic supply, fire, and emergency use. Includes how to make ferrocement water tanks.*
11. **Residential End Uses of Water Study (REUWS)** Peter W. Mayer, William B. DeOreo, Eva M. Opitz, Jack C. Kiefer, William Y. Davis, Benedykt Dziegielewski, and John Olaf Nelson, 1999. American Water Works Association/AWWA Research Foundation, 6666 W. Quincy Ave., Denver, CO 80235. Phone: 800-926-7337, 303-794-7711, Fax: 303-347-0804 www.awwa.org/bookstore *Considering the volume of water used in homes, we are amazingly ignorant of what it is used for. There are just a handful of studies. In this study special data loggers were attached to the water meters in 1,188 homes, in 14 cities. The data obtained from these devices allowed the researchers to separate water usage into individual end-use categories. A total of 6,000 households responded to a survey about their water use, and 12,000 households had their billing records incorporated into the study.*
12. **Unless you are in Texas,** *where use of greywater to intentionally keep expansive clay soils around a foundation permanently wet is officially encouraged. Opinions vary on the advisability of this setup. If your greywater system fails your foundation is liable to crack!*
13. **Cost/Benefit Spreadsheet** oasisdesign.net/downloads/rewatergwcalculator.xls *Example of a comprehensive cost/benefit analysis from ReWater Systems.*[s5]
14. **Complete Home Plumbing, 2nd edition** Scott Atkinson, editor, 2001. Sunset Books, Menlo Park, CA. www.sunset.com/sunset/bookstore *All the plumbing know-how you need for repairs, remodels, or even building from scratch can be found in this complete guide. Easy-to-follow text and photographs illustrate basic and advanced projects, from simply fixing a leaky faucet to remodeling a kitchen or adding a bath. There's also a chapter that covers plumbing outdoors. A must for every homeowner's bookshelf.*
15. **Principles of Ecological Design** Art Ludwig, 2003. Oasis Design oasisdesign.net/design/principles.htm *Principles for redesigning our way of life to live better with less resource use (description on inside back cover).*
16. **Ianto Evans** www.cobcottage.com *Thanks to Ianto for insights about Radical Plumbing.*
17. **Cleaners for Greywater Systems** University of Arizona, Office of Arid Lands Studies, 1955 E. 6th St., Tucson, AZ 85719. Phone: 520-621-1955, Fax: 520-621-3816. Contact: Martin Karpiscak, Phone: 520-621-8589, karpisca@ag.arizona.edu www.arid.arizona.edu *Independent lab assessment of cleaners for greywater; general principles and lab results only, no interpretation. This information is reproduced in our* Builder's Greywater Guide[6] *with some interpretation.*
18. **Untried ideas for urine management in dry, clay soil environments:** *1. A Constructed Wetland populated with plants that extract the nutrients but not the salt (assuming such plants exist). The leftover salty water could be routed to a disposal field, and the nutrient-laden wetland plants harvested and spread around the garden as fertilizing mulch.*
2. I suspect that salinization could be slowed in critical environments by using less dietary salt, perhaps by substituting potassium chloride instead of sodium chloride for seasoning.
19. **Cadillac Desert** Mark Reisner, 1986. Viking Penguin Inc., New York, NY us.penguingroup.com *Exceedingly well researched history of Western water.*
20. **Washington Toxics Coalition** www.watoxics.org *Alternatives to most household toxins.*
21. **The Book of Bamboo** David Farrelly, 1984 (reissued in 1995). Sierra Club Books www.sierraclub.org/books **Bamboo World** Victor Cusack, Deirdre Stewart, 2000. Simon & Schuster Australia/Kangaroo Press, Pymble, NSW, Australia. www.simonsays.com
22. **Designing and Maintaining Your Edible Landscape Naturally** Robert Kourik, 1986. www.robertkourik.com. Phone: 707-874-2606. *Landmark reference work on edible landscaping. A phenomenal quantity of quality information. Out of print, but available used .*
Edible landscaping is a huge topic. Ideally an edible landscape is designed in "four dimensions" (i.e., time as well as 3D space). Properly selected varieties yield a steady stream of fruit of all kinds all throughout the growing season (which could be all year), instead of an inundation of fruit for a few months. See our website for more information: oasisdesign.net/landscaping/fruittrees.htm
24. **Alternative Septic System: Watson Wick** oasisdesign.net/compostingtoilets/watsonwick.htm Tom Watson, PO Box 8, Embudo, NM 87531. Phone: 505-501-0949. *Available for consulting.*
25. **The Food and Heat Producing Solar Greenhouse: Design, Construction, Operation** Rick Fisher and Bill Yanda, 1980. John Muir Publications, Inc., PO Box 613, Santa Fe, NM 87504. Phone: 505-982-4078, 800-888-7504. *Excellent. Out of print but available used online.*
26. **Greywater for the Greenhouse** A.A. Rockefeller, 1979. Clivus Multrum Inc. www.clivusmultrum.com **The Greenhouse as Leachfield** A.A. Rockefeller, 1978. NSF International (formerly the National Sanitation Foundation), PO Box 130140, 789 N. Dixboro Rd., Ann Arbor, MI 48113-0140. Phone: 800-NSF-MARK, info@nsf.org www.nsf.org
27. **Design Manual: Constructed Wetlands for Municipal Wastewater Treatment** US Environmental Protection Agency, Office of Research and Development, 1988. Center for Environmental Research Information (CERI), Center for Environmental Research Information, 26 W. Martin Luther King Dr., Cincinnati, OH 45268. Phone: 513-569-7562, Fax: 513-569-7566 es.epa.gov/ncer/publications
28. **How to measure a gravity flow of greywater with a dipper box:** *Get the inlet set right so it cannot pin the dipping trough in the down position (adding a double ell to the inlet helps). Then there will be one dip for every 1½ gal of flow. With a battery-powered counter and a magnetic reed switch you can count dips, thus measuring an intermittent flow of cruddy water at atmospheric pressure. This is just about impossible to do any other way.*
29. **Can a 4,000 ft² Home Be Green?** oasisdesign.net/faq/green4000ft2home.htm
30. **Monitoring Greywater Use: Three Case Studies in California** California Department of Water Resources, Publications and Paperwork Management Office, PO Box 942836, Sacramento, CA 94236-0001. Phone: 916-653-1097.

Contact: Susan Sims, Phone: 916-653-6192, Fax: 916-653-4684 www.publicaffairs.water.ca.gov/information/pubs.cfm *Also available from* oasisdesign.net/greywater/SBebmudGWstudy.htm

[31] **Dry Toilets** Centro de Innovación en Tecnología Alternativa, A.C. Ave. San Diego No. 501, Col. Vista Hermosa, C.P. 62290, Cuernavaca, Morelos, Mexico. Phone: 52-777-322-8638 acua@terra.com.mx www.laneta.apc.org/esac/citaesp.htm *Well-proven designs for dry toilets.*

[33] **Greywater Heat Recovery** NATAS (National Appropriate Technology Assistance Service), PO Box 3838, Butte, MT 59702-2525. Phone: 800-275-6228 www.ncat.org

[34] **Indoor Greywater Reuse** oasisdesign.net/greywater/indoors *Indoors, the "receiving landscape" is plumbing fixtures rather than plants and soil. Especially noteworthy is the Toronto Healthy House, where all greywater is recycled directly to fixtures.*

[35] **Residential Water Reuse** Murray Milne, 1979. Report #46. Regents of the University of California. California Water Resources Center, University of California at Davis. University of California Center for Water Resources, Rubidoux Hall 094, University of California, Riverside, CA 92521-0436. Phone: 951-827-4327, Fax: 951-827-5295 www.waterresources.ucr.edu. *Out of print but available from* National Technical Information Service (NTIS), 5285 Port Royal Rd., Springfield, VA 22161. Phone: 800-553-6847, 703-605-6000 www.ntis.gov *550 pages, highly readable goldmine of fascinating information and history of water reuse. Includes 50 page annotated bibliography.*

[36] **Excreta Disposal for Rural Areas and Small Communities** E. Wagner and J. Laniox, 1958 (reprinted 1971). Monograph #39, World Health Organization, Geneva, Switzerland www.who.int/publications/en *Information on the effectiveness of soils for containing and processing nutrient and pathogen contamination from human feces, as well as other technologies.*

[37] **Composting Toilet System Book** David Del Porto and Carol Steinfeld, 1999. Ecowaters Projects, PO Box 1330, Concord, MA 01742. Phone: 978-318-7033 www.ecowaters.org *Tons of raw information on composting toilets.*

[38] **The Toilet Papers: Recycling Waste and Conserving Water** Sim Van Der Ryn, 1978. Reprinted 1999 by Chelsea Green Publishing. 85 N. Main St., Ste. 120, White River Jct., VT 05001. Phone: 800-639-4099, 802-295-6300, Fax: 802-295-6444 www.chelseagreen.com

[39] **Toilet Alternatives** Art Ludwig (book in progress) oasisdesign.net/compostingtoilets/book *Comprehensive guide to alternatives to the flush toilet for homeowners, builders, and regulators.*

[40] **Rainwater Harvesting and Runoff Management** Art Ludwig (book in progress) oasisdesign.net/water/rainharvesting

[41] **Long-Term Effects of Landscape Irrigation Using Household Greywater – Literature Review and Synthesis** L. Roesner, Y. Qian, M. Criswell, M. Stromberger, S. Klein. 2006. Water Environment Research Foundation and the Soap and Detergent Association. *Excellent summary of current state of greywater science. Full study available free in PDF format from www.werf.org.*

Suppliers (References s1-)

Caution: Unless specifically noted, we don't know anything about the suppliers' experience or the quality of their offerings. Supplier information changes rapidly; this list was last updated February 2009.

[s1] **North American Salt Co.** 9900 W. 109th St., Ste. 600, Overland Park, KS 66210. Phone: 877-IMC-SALT, Fax: 877-423-7258. Contact: Jerry Poe, Phone: 913-344-9195, Fax: 913-338-7924, poe@imcsalt.com www.nasalt.com *Manufacturer of potassium chloride water softener salt.*

[s2] **Jandy Products** PO Box 6000, Petaluma, CA 94955. Phone: 800-227-1442, 707-776-8200, Fax: 800-526-3928, 707-763-7785, info@jandy.com www.jandy.com *Makers of 3-way diverter valves. You'll more likely find these through pool and spa places than regular plumbing supply houses. As a convenience, we also sell these valves through* oasisdesign.net/catalog

[s3] **Orenco Systems** 814 Airway Ave., Sutherlin, OR 97479-9012. Phone: 800-348-9843, 541-459-4449, Fax: 541-459-6781 www.orenco.com *Established maker of residential and small community secondary treatment systems, sells components, gives seminars.*

[s4] **Fluid Dynamic Siphons, Inc.** PO Box 882019, Steamboat Springs, CO 80488-2019. Phone: 800-888-5653, 970-879-2494, Fax: 970-879-4948, info@siphons.com www.siphons.com

[s5] **ReWater Systems** PO Box 210171, Chula Vista, CA 91921. Phone: 619-421-1921. Contact: Steve Bilson www.rewater.com *Maker of plastic distribution cones and a range of GW systems from $1,295 and up (2007). Source for Zabel effluent pumps and burial rated pump tanks. Active in greywater politics.*

[s6] **Pacific Echo, Inc.** 23540 Telo Ave., Torrance, CA 90505. Phone: 310-539-1822, 800-421-5196, Fax: 310-539-5826 www.pacificecho.com *Makers of flexible PVC. Call for local distributor.*

[s7] **Real Goods** 833 W. South Boulder Rd., Louisville, CO 80027. 800-919-2400, 303-222-3600, Fax: 800-508-2342 www.realgoods.com *Mail order source for spa-flex, 3-way diverter valves, the "lid" sink, and the Earthstar Greywater System (a copy of the old AGWA system). There haven't been that many Earthstar systems sold, and there isn't much data available to say how they are working out in the field—they may be really good.*

[s8] **Little Giant Pump Company** PO Box 12010, Oklahoma City, OK 73157-2010. Phone: 888-956-0000, 405-947-2511, Fax: 405-947-8720, CustomerService@LittleGiant.com www.lgpc.com

[s9] **Oasis Design** Santa Barbara, CA. Phone: 805-967-9956, Fax: 805-967-3229 oasisdesign.net/catalog, oasisdesign.net/about/contact. *Supplier of double ells pre-drilled with access plugs, 3-way diverter valves, fruit tree and plant lists, books (including this one).*

[s10] **Geoflow** 506 Tamal Plaza, Corte Madera, CA 94925. Phone: 800-828-3388, 415–927–6000, Fax: 415-927-0120 www.geoflow.com Contact: Karen Ferguson. *Their premium underground drip irrigation tubing is impregnated with herbicide to keep roots out and has a good reliability record.*

[s11] **Oasis Biocompatible Cleaners** Distributed by Bio Pac, Inc., 584 Pinto Court, Incline Village, NV 89451. Phone: 800-225-2855, 775-831-9493, 702-425-5492, Fax: 866-628-1662, info@bio-pac.com Contact: Collin Harris www.bio-pac.com *Makers of plant and soil biocompatible cleaners.*

[s12] **Aqua-Flo Supply** 30 S. La Patera Ln., Unit 10, Goleta, CA 93117. Phone: 805-967-2374, Fax: 805-967-5509. www.aquaflo.com *A storefront/UPS source for greywater system components.*

[s13] **Infiltrator Systems** 6 Business Park Road, PO Box 768, Old Saybrook, CT 06475. Phone: 800-718-2754, 860-577-7000, Fax: 860-577-7001, info@infiltratorsystems.com www.infiltratorsystems.com *Plastic "infiltrator" gravel-less infiltration chambers of good quality.*

[s14] **NutriCycle Systems** (formerly Hanson Associates) 3205 Poffenberger Rd., Jefferson, MD 21755. Phone: 301-371-9172. Contact: John Hanson, jhanson@nutricyclesystems.com www.nutricyclesystems.com *"Nutrient recycling system" composting toilet, box trough greywater systems. Leaching chamber and box trough designs are thanks to John Hanson.*

[s15] **Clivus Multrum Inc.** 15 Union St., Lawrence, MA 01840. Phone: 800-4-CLIVUS (800-425-4887), 978-725-5591, Fax: 978-557-9658 www.clivusmultrum.com *Longtime manufacturer and distributor of composting toilets. Supplier of greywater systems and information. Half-pipe infiltration galleys, greenhouse, and freeze switch designs are courtesy of Carl Lindstrom.*

[s16] **John Todd Ecological Design, Inc.** PO Box 497, Woods Hole, MA 02543. Phone: 508-548-2545, Fax: 508-540-3962, info@toddecological.com www.toddecological.com *Pioneer provider of site specific, integrated ecological wastewater solutions including greenhouse aquatics, Constructed Wetlands, stormwater management, etc.*

[s17] **Tad Montgomery & Associates** 118 Washington St. #2, Brattleboro, VT 05301. Phone/Fax: 802-251-0502, eco@tadmontgomery.com tadmontgomery.com *Ecological engineering and design, focus on code-approved systems in northeast US, Constructed Wetlands, composting toilets, etc.*

[s18] **Natural Systems** 3600 Cerrillos Rd., Ste. 1102, Santa Fe, NM 87507. Phone: 505-988-7453, Fax: 505-988-3720, nsi@natsys-inc.com Contact: Michael Ogden www.natsys-inc.com *One of the first engineering firms specializing in Constructed Wetlands—a good outfit.*

[s19] **Northwest Water Source** (formerly Greywater Management) PO Box 2766, Friday Harbor, WA 98250. Phone: 360-378-8900, Fax: 360-378-8790. Contact: Tim Pope, water@waterstore.com. *"Aquabank" auto-backwashing sand filter and ozone/UV disinfection system. Alternative water sourcing, rainfall catchment.*

[s20] **Polylok Inc.** 3 Fairfield Blvd.,.Phone: 888-POLYLOK (888-765-9565), Fax: 203-284-8514 www.polylok.com *Call them to find local dealers, then ask local dealers about making a custom dipper box for you, modify a standard dipper, or use the components to make one yourself.*

[s21] **Homestead Utilities** 17366 E. Meadow Ln., Mayer, AZ 86333-4119. Phone: 800 CYCLE-H2O (800-29253-426), Fax: 928-632-9114. Contact: Anton, theo@leque.com www.leque.com (click on Products) *According to their literature, makes a $556 (2005) system that filters and chlorinates greywater to reuse for flushing toilets.*

[s22] **WaterFilm Energy Inc.** PO Box 128, Medford, NY 11763. Phone: 631-758-6271, Fax: 631-730-3918, info@gfxtechnology.com www.gfxtechnology.com *Makers of a greywater heat exchanger. It likes ≥24" fall to do its job, and works best with extensive shower use.*

[s23] **Valterra Products, Inc.** 15230 San Fernando Mission Blvd., Ste. 107, Mission Hills, CA 91345. Phone: 818-898-1671, Fax: 818-361-5389 www.valterra.com *Makers of RV dump valves.*

Index